LET'S MAKE THINGS GO!

ALL ABOUT ENGINES
for Young Scientists

LET'S MAKE THINGS GO!

ALL ABOUT ENGINES
for Young Scientists

Published by
Heron Books, Inc.
20950 SW Rock Creek Road
Sheridan, OR 97378

heronbooks.com

Special thanks to all the teachers and students who provided feedback instrumental to this edition.

Third Edition © 1976, 2022 Heron Books.
All Rights Reserved

ISBN: 978-0-89-739249-5

Any unauthorized copying, translation, duplication or distribution, in whole or in part, by any means, including electronic copying, storage or transmission, is a violation of applicable laws.

The Heron Books name and the heron bird symbol are registered trademarks of Delphi Schools, Inc.

Printed in the USA

18 July 2022

At Heron Books, we think learning should be engaging and fun. It should be hands-on and allow students to move at their own pace.

To facilitate this we have created a learning guide that will help any student progress through this book, chapter by chapter, with confidence and interest.

Get learning guides at
heronbooks.com/learningguides.

For teacher resources,
such as a final exam, email
teacherresources@heronbooks.com.

We would love to hear from you!
Email us at *feedback@heronbooks.com.*

IN THIS BOOK

1 MACHINES — 1

2 FORCE — 2
    Force and Motion — 2
    Let's Do This: Force — 6
    Let's Do This: Windmill — 8

3 FRICTION — 12
    What Causes Friction? — 13
    More and Less Friction — 14
    Let's Do This: Friction — 17

4 GRAVITY — 18
    Let's Do This: Gravity — 22

5 WORK AND ENERGY — 24
    Work — 24
    Energy — 26
    Heat Energy — 28
    Light Energy — 30
    Electrical Energy — 30
    Motion Energy — 32

6 STORED ENERGY — 34
    Let's Do This: Paper Airplane Launcher — 36

7	ENERGY CHANGES	38
	Changing Where the Energy Is	39
	Changing What Kind of Energy It Is	40
	Changing to Stored Energy	41
	Let's Do This: Spool Racer	42

8	FUEL	44

9	CHANGING FORCES	48
	Changing Forces More Than Once	48

10	ENGINES	50
	Engines that Use Motion Energy	52
	Engines that Use Stored Energy	52
	Let's Do This: Rubber Band Helicopter	54

11	ROCKETS AND JETS	58
	A Simple Jet Engine	59
	Rocket Engines	60
	Jet Airplane Engines	62
	Let's Do This: Balloon Jet Engine	64
	Let's Do This: Bubble Jet Engine	66
	Let's Do This: Match Rocket Engine	68

12	ELECTRIC MOTORS	70
	Let's Do This: Table Fan	72

13	SOME INTERESTING ENGINES AND MACHINES	74
	Mars Rover	76
	Let's Do This: Pulse Jet Engine	78

CHAPTER 1 MACHINES

The world is filled with machines that help us get things done.

A **machine** is a piece of equipment used to make work easier. Think about blenders, electric toothbrushes and electric ice cream makers. The purpose of any machine, whether simple or complicated, is to make work easier.

A computer is a machine.

So is a bicycle, and a toaster.

People use sewing machines to make clothes.

Your dentist uses machines to care for your teeth.

We use printers to print photos and documents.

We clean carpets with vacuums and use lawn mowers to cut grass.

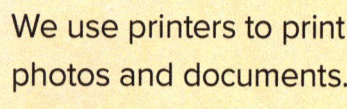

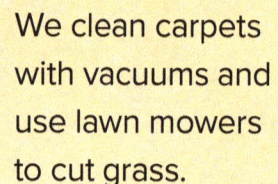

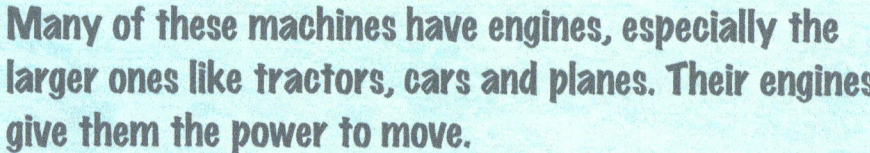

Many of these machines have engines, especially the larger ones like tractors, cars and planes. Their engines give them the power to move.

An **engine** is a special kind of machine used to make things go.

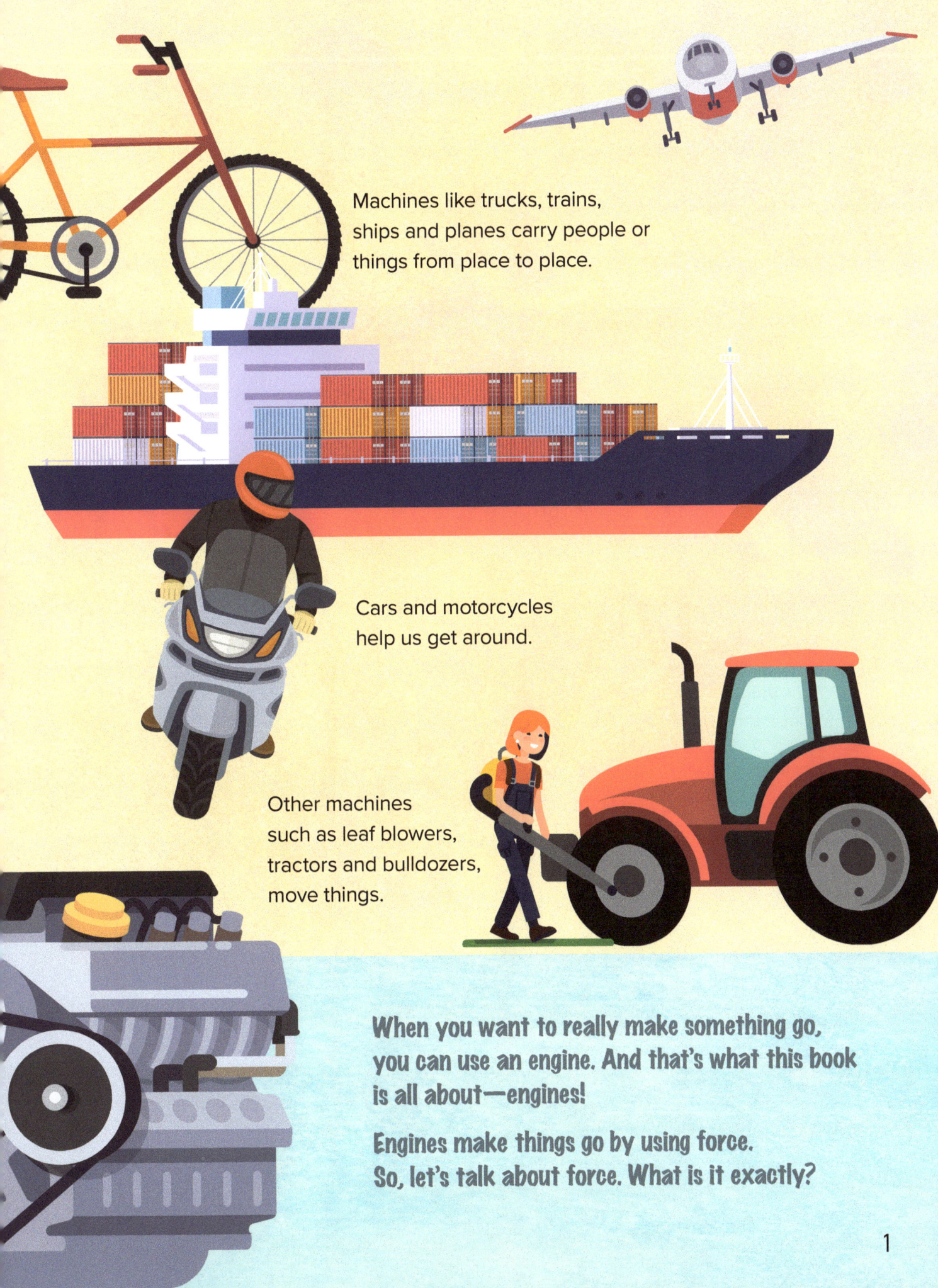

Machines like trucks, trains, ships and planes carry people or things from place to place.

Cars and motorcycles help us get around.

Other machines such as leaf blowers, tractors and bulldozers, move things.

When you want to really make something go, you can use an engine. And that's what this book is all about—engines!

Engines make things go by using force. So, let's talk about force. What is it exactly?

CHAPTER 2 FORCE

In science, a **force** is a push or a pull, no matter how weak or strong. A force can be small enough to lift a piece of dust, or strong enough to knock down a tree. Either way, it's a force.

FORCE AND MOTION

Starting and Stopping

Not every object needs the same amount of force to get it moving.

You know from experience that it takes more force to move something heavy than to move something light. You know that it takes more force to move a large rock than it does to move a balloon the same size.

The same is true of slowing down or stopping the rock and balloon. Once they're moving, they both want to keep moving.

The force needed to slow down or stop the rock is much greater than the force needed to slow down or stop the balloon.

When an object is not moving, it stays still until a force gets it moving. In other words, once you stop your skateboard, it will just sit there until you get it moving again.

The skateboard needs something to get it moving. It needs some kind of force to give it a push or pull.

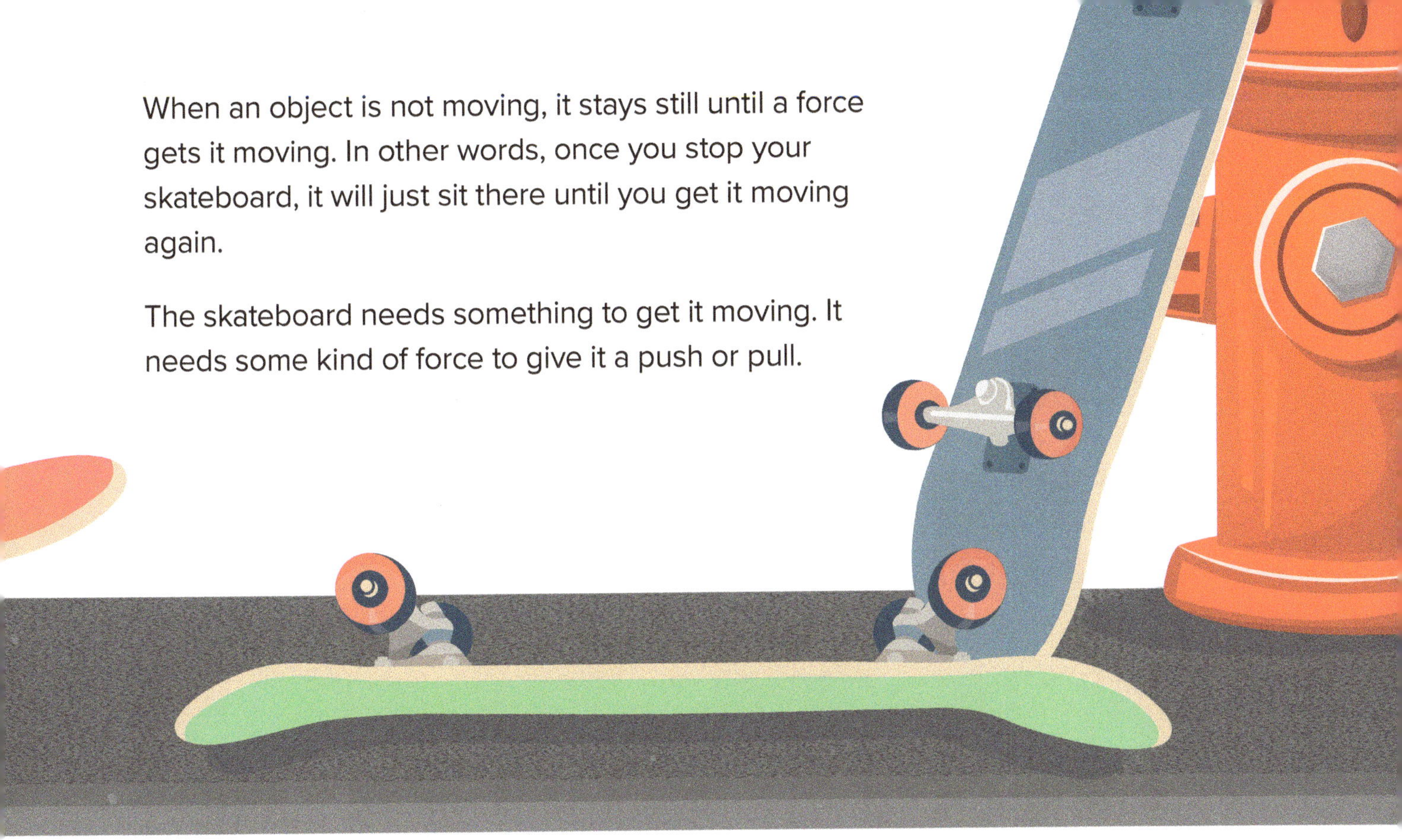

It takes force to get something moving. And it takes force to stop something.

Imagine trying to stop a rolling car using the same amount of force you would use for a skateboard. You would just be run over! There is a huge difference in weight between the skateboard and the car.

Starting and stopping heavy objects takes more force.

Starting and stopping lighter objects takes less force.

Speeding up and Slowing Down

Sometimes force isn't just used to start or stop objects. Force can also be used to speed up an object or slow it down

If something is moving and you add more force in the same direction it's going, it will move faster.

This is what happens when you're riding your bike and you want to speed up. You pedal harder, with more force, and your bike goes faster.

You can slow something down by using force in the opposite direction.

Changing Direction

A force can also change the direction something is moving. If a soccer ball is rolling toward the goal, the force of a kick can get it going in a different direction.

When you ride a bike, you use force to turn the handlebars and the bike changes direction.

So, force is used to start or stop objects, to slow them down or speed them up, or to change their direction.

force →

direction →

Wind blows and sailboats move across the water.

When you ice skate, your skate pushes against the ice and you move.

direction →

← **force**

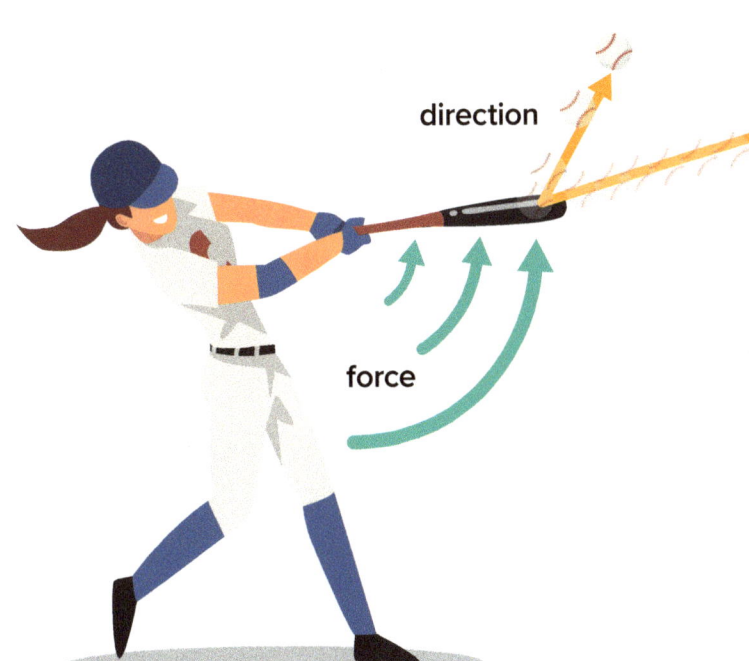

direction

force

You swing a baseball bat. The force of your swing changes the direction of the ball, pushes it into the air and, hopefully, over the fence for a home run!

If you stop and look around, you will see forces at work everywhere.

FORCE

For this activity you will need

- pencil
- various other objects

Steps

1. Pick out three things in the room that are not moving and weigh different amounts.

2. Use force to move each of those things. Notice the difference in the force you need to use.

3. Pick out three things in the room that are moving (or you could get moving), and weigh different amounts.

4. Once it's moving, use force to stop each of those things. Notice the difference in the force you need to use.

5. Get a pencil and another object, and set the object down on the table about a foot away from you.

6. See how much force is needed to roll the pencil until it just hits the object.

7. Move the object further away, and see how much force is needed to roll the pencil to it.

8. Move the object closer to you and see how much force is needed to roll the pencil to it.

9. Tell (or write up for) someone what force is in science.

WINDMILL

For this activity you will need

- paper
- scissors
- hole punch
- plastic straw
- modeling clay
- thread or string 2 feet long
- 15-20 regular-sized paper clips
- tape

Steps

1. Trace the windmill pattern on a sheet of paper. Cut out the square.

2. Using the hole punch, punch out the black dot in each corner.

3. Cut along each dashed line of the square from the corner up to the circle.

4. Use the pointed end of a pencil to make a hole in the center of the circle at the + sign. Make the hole just big enough to push a straw through it.

5. Push the straw through the center of the circle so that the end of the straw sticks through about an inch.

6. Bend one corner of the square over the end of the straw, and push the straw through the hole in it. Do the same with the other corners. Now you have your windmill blades.

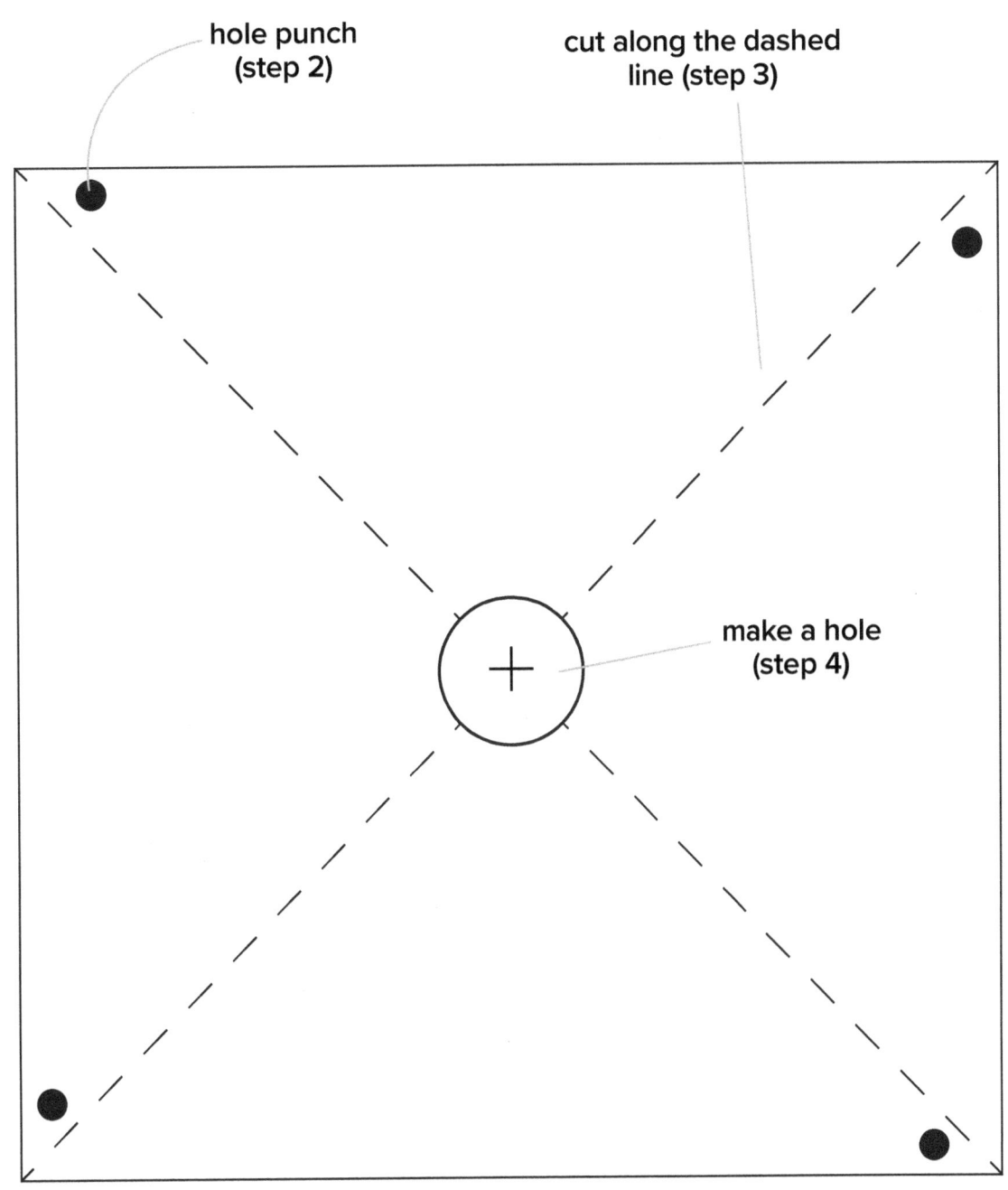

WINDMILL PATTERN

7 Slide the blades to the middle of the straw. Hold the blades in place on the straw with modeling clay.

8 Use tape to attach one end of the sewing thread near one end of the straw.

9 Tie a regular-sized paper clip onto the free end of the thread.

10 Hold the windmill loosely between your thumb and finger as shown in the picture and blow on the blades. Be sure to hold the straw loosely so it can turn freely. Can you lift the paper clip with the force of your breath?

tape attached to thread (step 8)

11 Based on the experiment you have done so far with lifting a paper clip with your breath, imagine how many paper clips you think you will be able to lift.

12 Now see if your breath and windmill can lift 5 paper clips. (You can clip 4 more paper clips onto the one you have, or you can clip them together in a chain.) Keep adding paper clips to see how many you can lift and how close your guess was.

13 (Optional) Based on what you've done so far, think about how you might be able to make your windmill able to lift even more paper clips. If you think you have a good idea, try it out with a new experiment.

14 Tell (or write up for) someone what you did and what you observed. Be sure to include why your windmill is a machine and how your breath provided the force to lift the paper clips.

CHAPTER 3 FRICTION

Imagine you have a book sitting in front of you and you want to give it to a friend on the other side of the table. You give it a quick push and it slides across.

Now imagine you're sitting on a carpeted floor with the book in front of you. Again, you want to slide it over to someone so you give it the same quick push. You use the same amount of force as before, but the book hardly moves.

What made the difference? Friction.

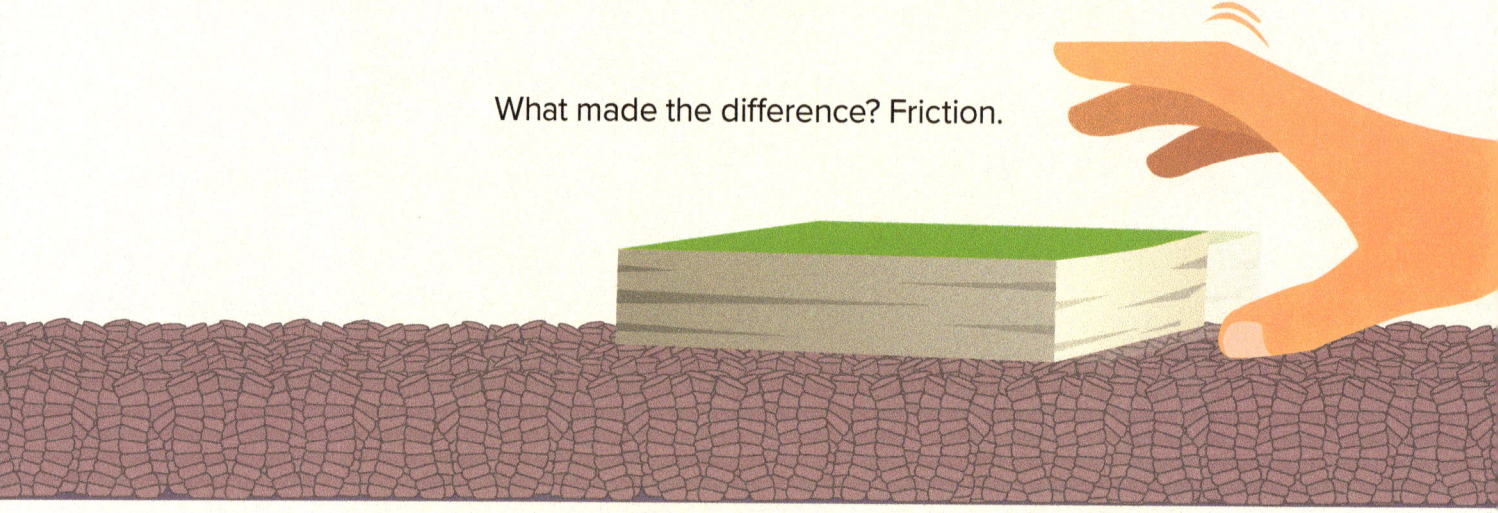

Friction is a force that happens when one surface rubs against another. Friction slows down or stops things. A tabletop is fairly smooth. There is little friction, so the book slides easily. But the carpet is not so smooth, so the book doesn't slide as easily. There was more friction.

WHAT CAUSES FRICTION?

In the examples of the carpet and the table, it's easy to see that a rough surface has more friction than a smooth surface. But let's look at this more closely.

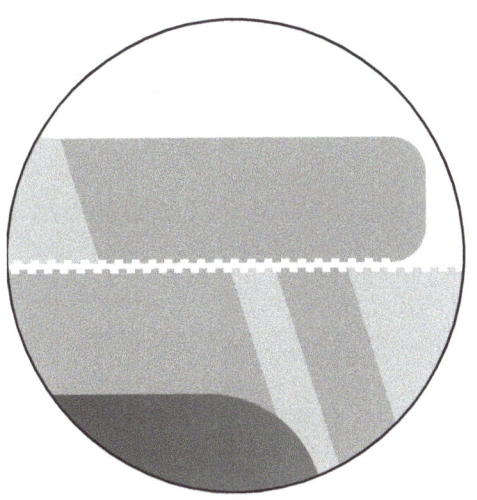

The surfaces of objects are not perfectly smooth, even when they appear to be.

They have tiny ridges and bumps. When two objects rub against each other, the tiny ridges and bumps on one object push against the ridges and bumps on the other object.

That rubbing force is friction. It's a force that slows or stops motion.

When you roll a ball across a lawn, there is friction between the ball and the grass. The ball rolls for a bit, then slows and stops.

A racing bike with thin tires will roll for a long distance on a very smooth road. But it too will eventually stop.

There is friction between the wheels of a moving bicycle and the road. Even on a good, strong bike with tires full of air, if you stop pedaling on a flat road, the friction caused by the tires rubbing against the road will slow your bike down and eventually stop it.

Friction between your shoes and the sidewalk keeps your shoes from slipping and sliding.

When you walk down a sidewalk, there's usually enough friction for you to walk easily without slipping. When you take a step, the bottom of your shoe rubs against the sidewalk. This friction keeps your shoe from sliding, and you move forward.

MORE AND LESS FRICTION

But what if there is too much friction?

Too much friction slows motion, or might even stop it. Imagine you're riding a bike on a smooth road and you get up some speed. Suddenly the road turns to rough dirt and gravel. Your bike slows down. It's harder to pedal.

Why? Now there is more friction between your tires and the rough road, too much to keep up your speed. Your bike is slowed down, and moving it along takes a lot more work.

What if there is too little friction?

Suppose you go out to take a walk on a very cold day, and the sidewalk is covered in ice. When you take a step, you find yourself slipping and sliding.

Why? You have no grip because there's so little friction between your shoes and the surface you're walking on!

14

Friction is all around us, yet we don't think much about it. But without friction, it would be hard to do almost anything!

Imagine if all floors were so slippery you kept falling down!

Imagine if doorknobs were so slippery you couldn't turn them or the handlebars of your bike were so slippery you couldn't hold on!

You need the right amount of friction for whatever you are doing.

If you want something to slow down or not slip, you want more friction.

If you want something to move fast, you don't want much friction.

Can you think of activities where more friction is good? Can you think of ones where less friction is good? It all depends on what you're trying to do.

FRICTION

For this activity you will need

- book
- 15 plastic straws
- small amount of modeling clay

Steps

1. Lay a book on a table. With one finger, give the book a push hard enough to move it. Notice what happens.

2. Stick a few tiny balls of clay to the back of the book. Then, using about the same force as you did before, give the book a push. Observe what happens.

3. Now, take the clay balls off the back of the book. Line up 15 plastic straws side-by-side on the table, about 1 inch apart. Put the book on the straws. Using about the same force you used on the earlier pushes, give the book a push. Notice how far the book moves.

4. (Optional) If you want, think of a different experiment with pushing the book and friction, and try it out.

5. Tell (or write up for) someone what you observed about friction and motion.

CHAPTER 4 GRAVITY

Remember that force is a push or a pull that starts, stops or changes the motion of something.

Here on Earth there is a pulling force at work all the time. You are very familiar with it. It's called gravity.

Gravity is the pulling force that holds us to the earth. It's what makes things fall if nothing is keeping them up. The force we call gravity pulls everything toward the center of the earth.

Just like other forces, the earth's gravity can change the way things are moving.

When you throw a ball straight up into the air, the force of your throw (push) makes it fly up. But then, as the force of gravity pulls on the ball, it slows down. Eventually it stops. As gravity continues to pull on it, the ball starts coming back down. The pull makes it go faster and faster as it falls.

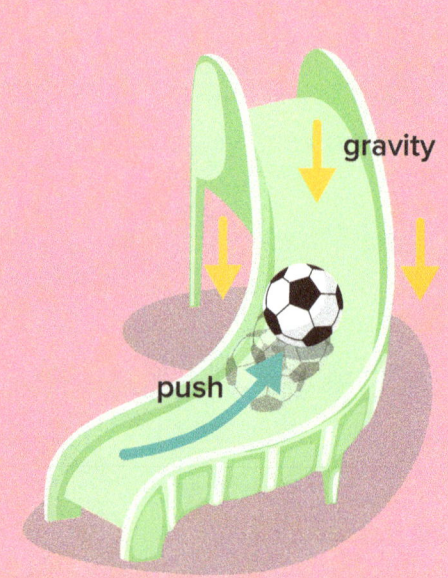

If you were to roll a ball up a playground slide, you would see the force of gravity working in the same way. The force of your push would make the ball roll up the slide for a while, but soon gravity would cause the ball to stop, then roll back down the slide.

18

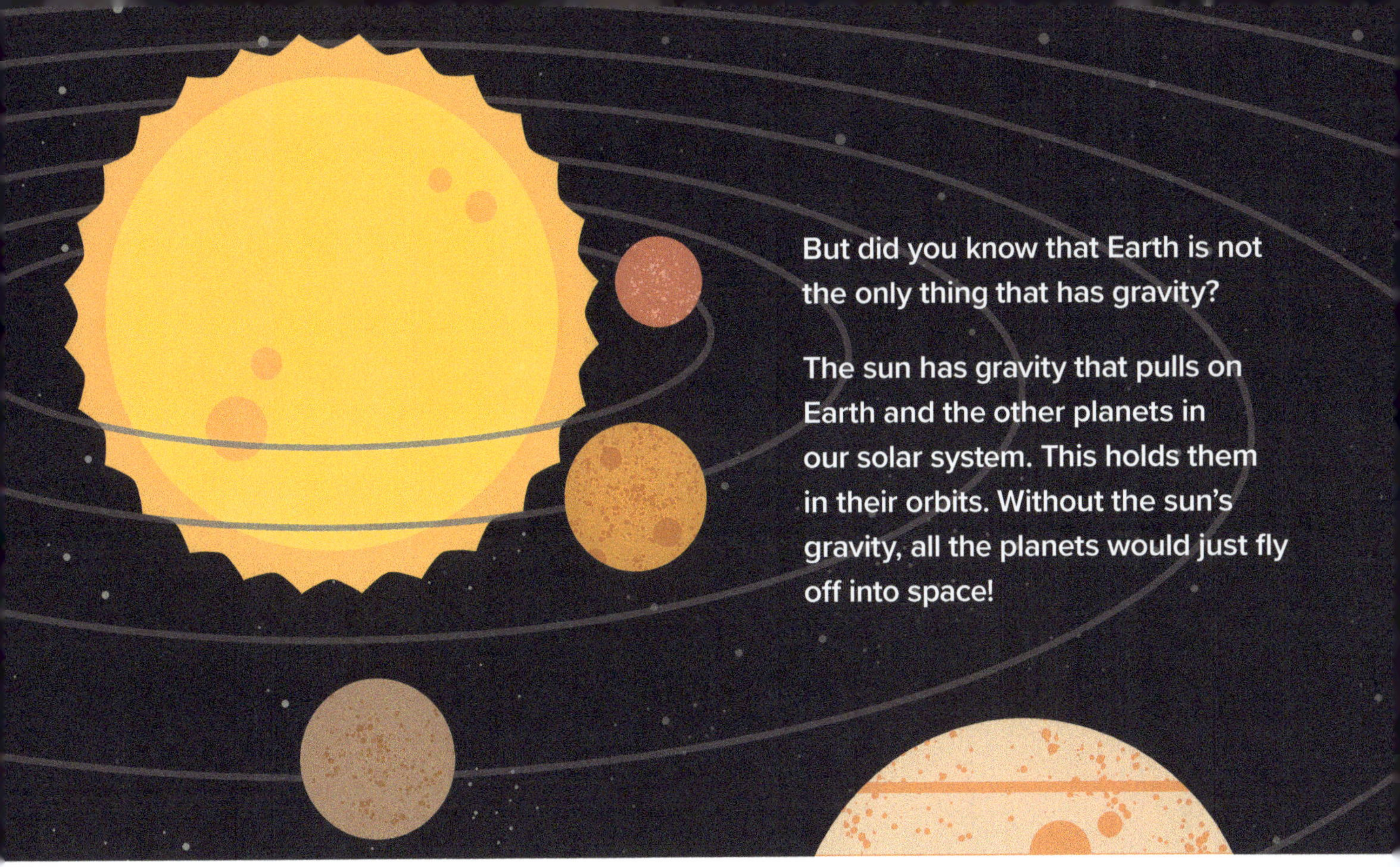

But did you know that Earth is not the only thing that has gravity?

The sun has gravity that pulls on Earth and the other planets in our solar system. This holds them in their orbits. Without the sun's gravity, all the planets would just fly off into space!

Have ever watched a rocket take off into space? If so, you may remember a huge burst of fire and smoke when the rocket was launched. At first, it seemed that the rocket was not moving at all. Then slowly it started to go up. Then it went faster and faster and finally escaped into space.

It takes a tremendous force to move something as heavy as a rocket off the surface of the earth. Gravity is trying to pull the rocket down, so the rocket must push away from Earth with a much stronger force.

Scientists figuring out how to get a rocket into outer space, and make it arrive to a distant place like the moon, have to understand a lot about the force of gravity. As a rocket moves farther and farther away from the earth, the force of gravity on the rocket becomes less and less.

But if the rocket is approaching the moon, then the gravity of the moon starts to pull on it. Because the moon is much smaller, its gravity is not as strong as Earth's. But it's large enough to create a pull you can feel.

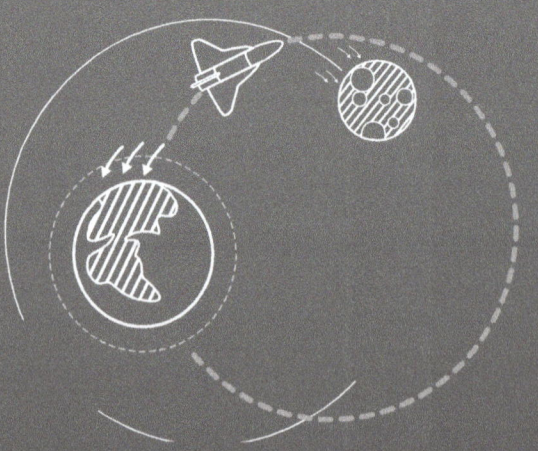

If you were to stand on the moon and throw a ball straight up, it would go almost 6 times higher than it would on Earth.

If you were to jump with the same amount of force you would use on Earth, you would rise 10 feet in the air!

On Earth, the pull of gravity is strong, and it's everywhere on the planet. It pulls on every object and every living thing all the time. If you've ever seen a very large object (such as a falling tree or building) come crashing to the earth, you were looking at the FORCE of gravity. It's strong!

Now, let's think again about getting a spaceship off the ground. Imagine the force needed to lift an elephant. That's a lot of force. Getting a rocket off the ground is like lifting 400 elephants!

Just imagine how powerful a rocket engine has to be to make that enormous rocket move off its launching pad and into outer space!

That's an engine that can make things go!

GRAVITY

For this activity you will need

- yard stick or meter stick
- string
- 15 paper clips

Steps

1. Cut three pieces of string about 2 feet long each.

2. Tie a paper clip to the end of each string.

3. Tie the strings to the yard stick or meter stick with one string at each end and one in the middle.

4. Hold the stick in the air so that the paper clips hang freely.

5. Slowly tilt the stick different directions and observe what the paper clips do.

6. Now add 4 paper clips to each of the first paper clips.

7. Again, slowly tilt the stick different directions and observe what the papers clips do.

8. Write or tell another person what you observed about the force of Earth's gravity.

5 CHAPTER WORK AND ENERGY

WORK

Normally, when a person says they are doing work, they mean they're doing their job, perhaps being a teacher or a nurse.

In science, "doing work" means something a little different. **Doing work** is using force, a push or a pull, to move something.

That may sound very simple, and it is. Any time something is pushed or pulled and it moves, work is done. It doesn't matter when or where. All that matters is that something is pushed or pulled and, as a result, it moves.

When the wind pushes a sailboat across a lake, the wind is doing work. It's moving the sailboat.

When a river floods and pulls a tree out of a riverbank and washes it downstream, the water is doing some work. It's moving the tree.

So "work" is moving something. This takes force.

Things look a little different when you think about work this way. For example, taking a minute to carry a stack of books from one side of your classroom to the other could actually be doing more of this kind of work than an hour or two of studying.

Why? Because carrying the books is using more force to move something, and that's doing more "work."

Which of these do you think is using more force to move something, and which is using less? Which one is doing more work?

In science, "doing work" is using force to move something.

ENERGY

Let's take a look at what energy has to do with doing work.

First, what do we mean by energy? Normally, when someone says they have "energy," it means they feel active and lively.

Scientists think about energy a little differently.

If something can move things, if it can do some work, we say it has energy. **Energy** is the ability to do work.

We see energy all around us whenever work is done. We see it when a car drives down the road.

We see it when we brush our teeth, plant a seed, paint a picture, shoot an arrow, bounce on a trampoline.

We see it when a boat is paddled down a stream, when a bike is ridden up a hill, when an airplane lifts off and soars up into the sky.

We see it when the wind moves clouds during a storm.

There are four common kinds of energy: heat energy, light energy, electrical energy and what we call motion energy. All these types of energy move things. They can all do work.

Let's look more closely at the four different kinds of energy.

HEAT ENERGY

To talk about heat energy, let's first talk about atoms.

As you may know, all the things around you (chairs, tables, pencils, books, food, cars, planes, clothes, even your body) are made of **atoms**. Atoms are so tiny you can't see them.

When we touch something that has fast-moving atoms, it feels warm or hot to us. The faster the atoms move, the warmer it feels.

Heat energy moves from warmer objects to cooler ones. When two things are close to one another, the fast-moving atoms of the warmer object make the atoms in the cooler one move faster and it gets warmer.

When you put a cool pan on a hot stove, heat energy from the stove makes the atoms in the pan start moving faster and it warms up. Now the pan has heat energy and it cooks your food.

Friction causes heat energy. If you rub your hands together, then touch your face, you can feel warmth from the friction.

The friction caused by striking a match creates so much heat the match catches fire.

An interesting thing about atoms is that they don't just sit still. They're always moving. Atoms jiggle around. Sometimes they jiggle more slowly and sometimes they jiggle faster.

The jiggling motion of atoms creates heat energy.

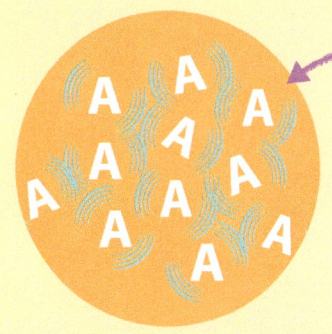

The faster they jiggle, the more heat energy there is.

And what about fire? Once a fire gets going, it gets very hot. A burning fire gives off a lot of heat energy.

Because atoms are so small, we can't really see heat energy at work, but we can certainly feel it. When you sit near a campfire, its heat energy warms you up. You might even get so hot you have to move away from the fire!

Think of a steaming cup of hot chocolate. Its atoms are moving fast. When you drink that hot chocolate on a cold day, you can feel its heat energy warming you up.

Heat energy helps us warm our homes, heat our water, cook our food, start useful fires, and dry our clothes. Heat energy is at work all around us every day.

LIGHT ENERGY

The sun, stars, lightning, light bulbs and lasers give off another kind of energy we call light energy. Light energy can also heat things up.

You might think it's the sun's heat that warms us here on Earth, but it's not. The sun is just too far away for that. It's the *light* from the sun that keeps the earth warm. Everything on Earth absorbs some light and reflects some light. Different objects do this in different amounts.

Light energy makes atoms move faster. When an object absorbs light from the sun, it heats up.

Maybe you've done the experiment of using a magnifying glass to focus light from the sun on a piece of paper. The magnifying glass concentrates so much light on one spot that the paper heats way up and catches fire. That's a good example of light energy at work.

ELECTRICAL ENERGY

Electrical energy is another type of energy that can do work. You see it everywhere—running washing machines, computers, toasters, lamps and televisions, for example.

Devices like these get their electrical energy from power stations that use power lines to send electricity to homes and workplaces.

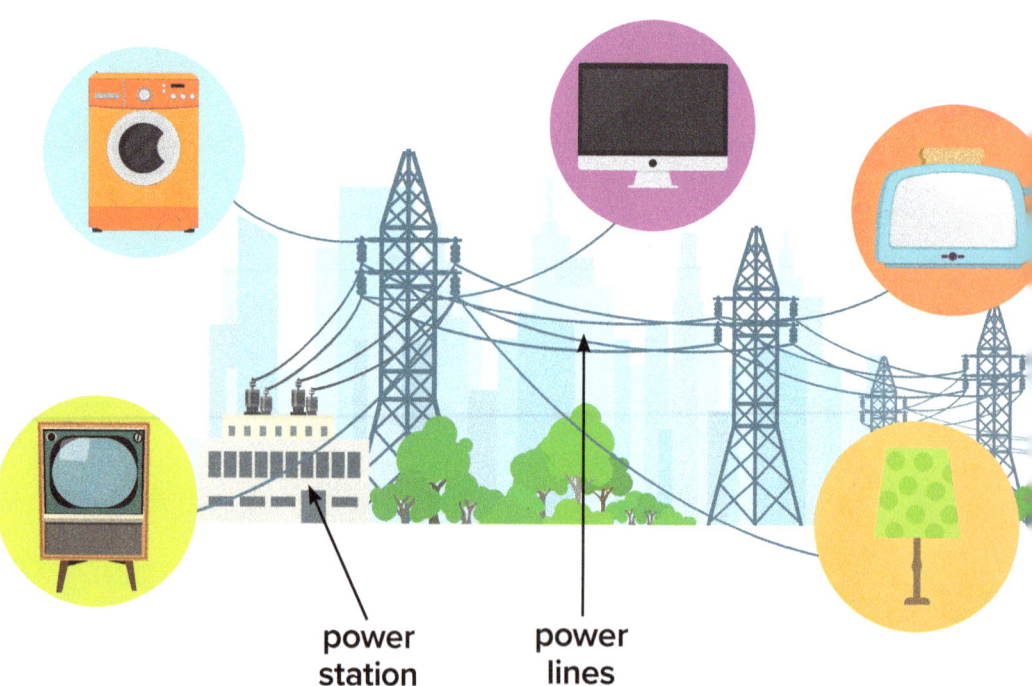

power station   power lines

Imagine you are taking a hike on a sunny day. You get tired and decide to sit down on a rock to rest. You might find the rock is too hot to sit on. Why? Some of the light from the sun has been absorbed into the rock and makes the atoms in the rock move around faster and faster. Remember, the rock gets hot from the light energy, not heat energy.

A similar rock sitting in the shade would not heat up like that, even if the air around the rock was hot.

Electrical energy also comes from the batteries found in watches, cell phones, flashlights, cameras and cars.

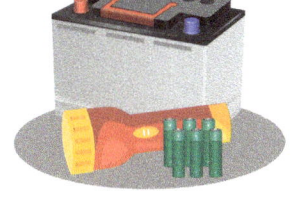

When you turn on an electrical device, whether it's a flashlight or a computer, tiny parts of atoms, called **electrons**, start moving. These moving electrons have electrical energy, and this is what lights up a flashlight, makes a washing machine run and makes a toaster get hot.

In larger machines like tv's and computers, these electrons travel through a wire plugged into the wall.

31

MOTION ENERGY

Things that are moving can make other things move by pushing or pulling them. Things that are moving have energy. Motion energy is the energy found in moving objects.

A hammer, for example, has motion energy when you swing it. When it strikes a nail, the nail pushes into the wall.

When you run at full speed and accidentally crash into someone, your motion energy can knock them over.

Moving air has motion energy.

A light breeze can make leaves move. Wind from a storm can break tree branches. A tornado can actually pick up a building, carry it a distance away, and put it down again.

Moving water has motion energy.

A rushing stream can push a boat along.

A bulldozer that is moving forward can push large piles of dirt around.

The stream of water from a hose can remove dirt from a car.

A wave can sweep someone off their feet or carry a surfer hundreds of yards.

6 CHAPTER STORED ENERGY

When you **store** something, you put it away to be used later. For example, you might store food in a refrigerator or cabinet.

Stored energy is energy that's being saved up so it can be used to do work and move things later on.

If you pull a rubber band back and let it go, it flies off. Where does it get the energy to do that? Let's take a look.

A rubber band gets some energy when you pull it back. It isn't using that energy yet, but at any moment it can. As long as it's pulled back, that energy is stored in the rubber band. It has stored energy.

If you pull a rubber band back, you are storing energy in it.

One kind of stored energy is the result of gravity. All objects higher than the ground have stored energy because when they're no longer held up, gravity will cause them to fall with motion energy.

A flying kite has stored energy. It's high up in the sky. The blowing wind keeps it in the air.

34

Then, when you let the rubber band go, it flies off! Why? It uses the stored energy to move.

And now it has motion energy.

Suddenly the wind dies down. What happens?

Gravity pulls it, and its stored energy becomes motion energy. The kite falls to the ground.

A battery is full of stored electrical energy. That energy can be used to make a flashlight shine or a cell phone work.

PAPER AIRPLANE LAUNCHER

For this activity you will need:

- sheet of 8 ½ by 11 inch paper (plain copier paper works well)
- rubber band – 3 ½ x ⅛ inch (size #33)
- hole punch

Steps

1. Fold the paper in half lengthwise. Then open it up.

   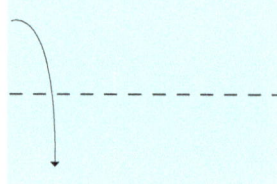

2. Fold the top two corners down to the center line. Run your fingernail down the folds to make them very sharp.

   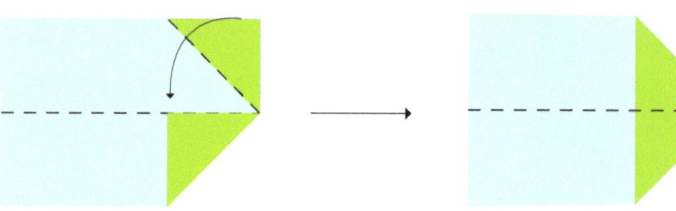

3. Fold the top folds down to the center line. Make the folds very sharp.

   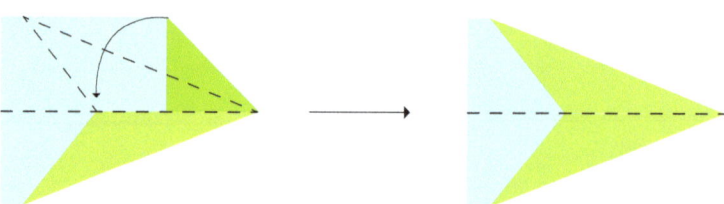

4. Fold the new top folds down to the center line, then fold them back up half way.

5️⃣ Your plane should look like this:

6️⃣ Use the hole punch to make a hole about 4 inches back from the pointed tip.

7️⃣ Thread your rubber band through the hole and pass one end of the rubber band through the loop at the other end of the rubber band.

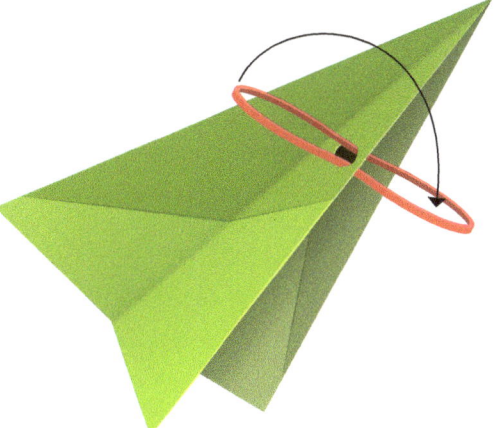

8️⃣ Hold your airplane with one hand where the rubber band is attached. Use the thumb of your other hand to stretch the rubber band out in front of you. Let go of the airplane and watch it fly.

9️⃣ Launch your airplane 3 or 4 more times. See if you can get it to go farther each time. Each time, write down what you did and what happened. Make sketches if you want to.

🔟 What do you think gravity and stored energy had to do with your results? Write down your thoughts.

CHAPTER 7 ENERGY CHANGES

We know that energy is used whenever force moves something and work is done.

What's interesting about this is that when energy does some work, it isn't used up or lost, it just changes.

Sometimes it's **where** it is that changes. It moves from one thing to another. For example, your moving foot kicks a soccer ball and the soccer ball flies off. Energy moves from your foot to the ball.

Sometimes the change is the **kind** of energy. It might change from electrical energy to heat energy, as happens when you put bread in the toaster and turn it on.

Sometimes energy is **stored**. It changes to stored energy that can be used later, as when you pull back on a rubber band. Your motion energy turns into stored energy.

38

CHANGING WHERE THE ENERGY IS

Let's look at more examples of energy moving from one thing to another.

A push is one way of moving energy from one thing to another. You use energy to push a skateboard. The skateboard moves. Now it has the energy of motion.

The energy changed where it was. It started with you, and now the skateboard has energy.

Another example is throwing a ball. You have motion energy in your arm. When you let go of the ball, the motion energy moves from your arm to the ball. And now the ball has motion energy.

Another example is jumping into a pool or pond. The motion energy of your body dropping into the water changes to motion energy of splashing water and waves moving away from you.

CHANGING WHAT KIND OF ENERGY IT IS

Let's look at more examples of changing the kind of energy.

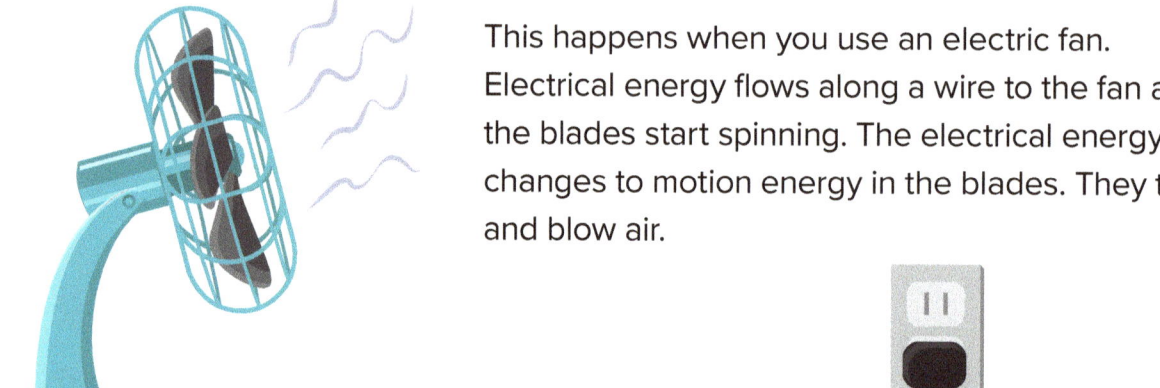

This happens when you use an electric fan. Electrical energy flows along a wire to the fan and the blades start spinning. The electrical energy changes to motion energy in the blades. They turn and blow air.

Energy changes into other kinds all the time.

When a hot air balloon rises in the sky, that's another example of energy changing from one kind to another. There's a burner right under the balloon that heats the air inside the balloon. Because hot air rises, the balloon goes up. Heat energy has changed to motion energy!

When you slam on your bicycle brakes, the motion energy of your bike tires skidding on the sidewalk turns into heat energy! For a few seconds, your tires are hot!

CHANGING TO STORED ENERGY

Some kinds of energy can be turned into stored energy for use later.

A simple wind-up toy, perhaps a little race car or a hopping frog, is an example.

You turn a knob on the outside of the toy. The motion energy of the turning winds a spring inside the toy until it is very tight. This puts energy into the spring. The spring stores it.

When you put the toy on the ground, the spring lets the stored energy out, a little at a time. This makes the toy move. The stored energy in the spring slowly turns back into motion energy and the car drives or the frog hops around.

Here's a fun example. Imagine jumping up and down on a trampoline.

Your motion energy down stretches the trampoline surface, storing energy in the trampoline.

Then the stored energy turns to motion energy and shoots you up into the air.

When you reach the top of your flight, you now have stored energy again. That stored energy turns into motion energy and down you go. You stretch the trampoline, creating stored energy again.

And all this repeats over and over. Motion energy to stored energy to motion energy to stored energy and on and on!

In all these examples, you can see that energy isn't getting lost or used up, it's just being changed in some way.

Energy is all around us, and it's changing all the time. It might be changing where it is, what kind of energy it is, or it might be getting stored up for future use.

41

Let's Do This!

SPOOL RACER

For this activity you will need

- thread spool
- rubber band slightly longer than the spool
- 2 round toothpicks
- 1 metal washer smaller than the end of the spool
- masking tape
- ballpoint pen

Steps

1. Insert the rubber band through the hole in the spool so that the loops hang out at both ends.

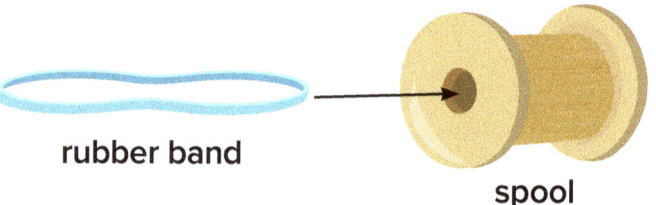

2. Put a toothpick through one loop. Tape the toothpick to the end of the spool on both sides of the rubber band.

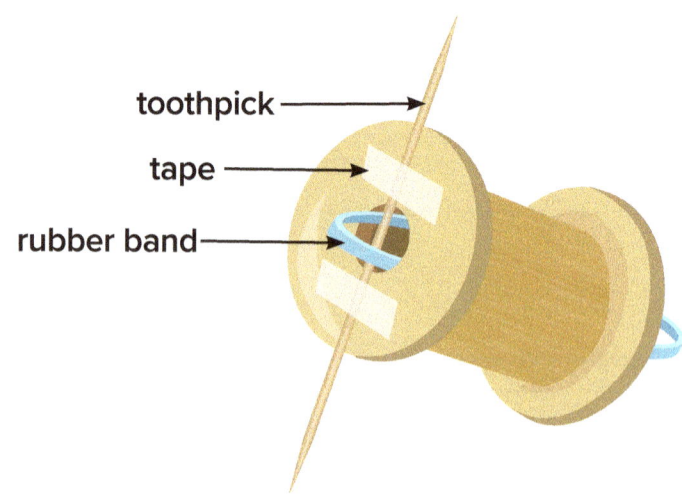

③ Break off the ends of the toothpick so they don't stick out over the edge of the spool.

toothpick

④ Thread the other loop of the rubber band through the washer.

washer

⑤ Put the ballpoint pen through this loop, but don't tape it down. The rubber band should be about ¼ of the way down the pen.

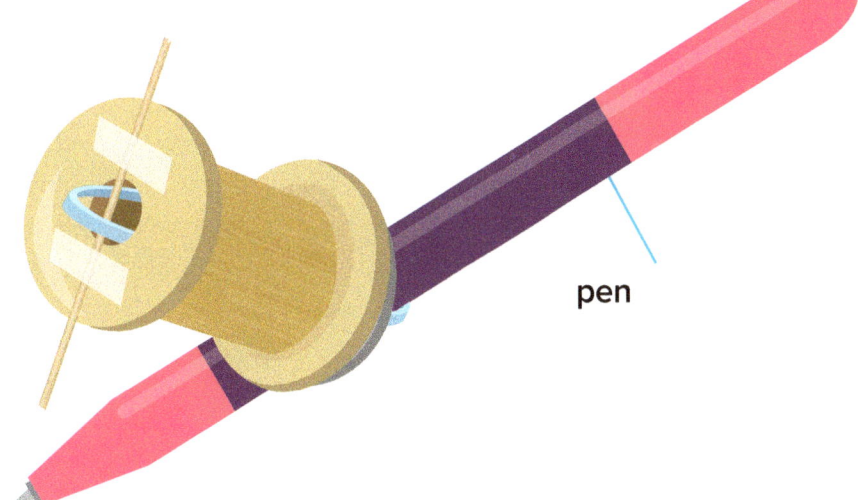
pen

⑥ Hold the spool in one hand and wind up the rubber band by turning the pen around and around.

⑦ Place the spool on a smooth, flat surface and let it go.

⑧ Tell (or write up for) another person what energy changes you observed, and what work was done (what force was used to move things).

CHAPTER 8
FUEL

A **fuel** is a material with stored energy that gets released when the material is burned. Coal, oil, gasoline and wood are common fuels with stored energy that is easily used.

When fuels are burned, their stored energy is released as heat energy and light energy. This can be used to do work.

But where do fuels get their stored energy?

Would you believe it comes from the sun?

Wood is a fuel we get from trees that live and grow in sunlight. Energy from the sunlight is stored in the wood. When wood is burned, this energy is released as heat energy and light energy.

Another fuel is gasoline made from oil that comes out of the ground. Gasoline has a lot of stored energy, and we use it to power things like cars, ships, lawn mowers, tractors, bulldozers and construction equipment.

1

The energy in oil and gasoline comes from plants and animals that lived and grew in sunlight long, long ago. These plants and animals stored energy from the sun.

Over millions of years, many, many of them lived and died. When they died, their bodies sank to the forest floor or the bottom of the sea. As this happened over millions of years, the layers of plant and animal bodies got deeper and deeper, and heavier and heavier.

2

As more and more layers built up and were buried in the earth, the weight of it all squeezed the layers together and this heated them up. Very slowly, over a long period of time, they changed into oil.

We also get coal from plants and animals that were buried long ago. Coal is similar to oil, but it has become a hardened solid rather than a liquid.

Coal, oil and gasoline are called **fossil fuels** because they come from the remains of plants and animals, called fossils, that lived long, long ago.

③ We get this oil out of the ground, clean it and turn it into gasoline.

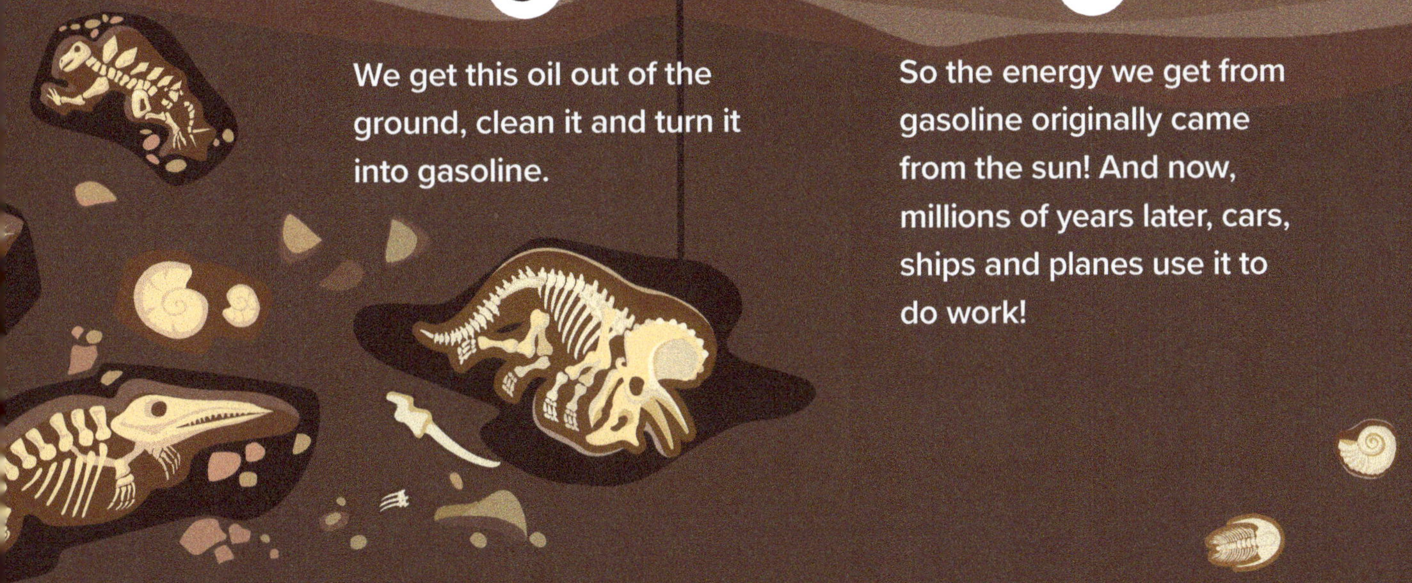

④ So the energy we get from gasoline originally came from the sun! And now, millions of years later, cars, ships and planes use it to do work!

Fossil fuels, especially gasoline, are used in many kinds of engines to make them go.

9 CHAPTER CHANGING FORCES

Machines make work easier by changing forces. A machine can change the amount of force needed to do something or it can change the direction of a force.

Let's say you want to lift a rock. The problem is that it's too heavy. There's a simple machine for this, called a crowbar, that can help you lift the rock. A crowbar changes the amount of force needed to move a large object.

Wheels are part of many machines. These also help lessen the amount of force needed to move things.

CHANGING FORCES MORE THAN ONCE

Machines that have many parts can change the direction or amount of force several times.

An example of a machine like this is a bicycle. When you push on the pedals, the force doesn't go straight from your feet to the tire. The force of your pedaling is changed in amount or direction several times.

Your feet move the pedals. The pedals move the gear they're attached to. A **gear** is a special kind of wheel with teeth around the edge.

The gear moves the chain. The chain moves your back wheel. At each of these steps, the amount or direction of your original force is changed.

If you wanted to move a heavy box, for example, trying to drag it along the ground would be hard. But putting it on a wagon or cart would change the amount of force you needed. You could just roll the cart along instead of having to drag the box.

A pulley is a simple machine that can help you raise and lower things by changing the direction of a force. An example is raising a flag up a large flagpole. When you use a pulley and a rope, you pull down on the rope and the flag goes up the pole.

People who design machines that change the amount or direction of forces are called **engineers**. Figuring out how to make machines that make different kinds of work easier is a fun challenge.

Engineers are always looking and learning more about all the things we've talked about in this book—machines, force, friction, gravity, work, different types of energy, fuel and changing forces.

Now, let's look at something else that engineers are often interested in, the thing their name (engineer) comes from, the thing that makes things go—engines!

49

10 CHAPTER ENGINES

All machines make work easier, but some machines do more than that. An **engine** is a special kind of machine that uses energy to make things go.

Car engines, for example, use energy to turn wheels. The tires push against the road and the car moves forward.

Rockets have engines that shoot them into outer space.

Boats have engines that turn propellers that move them through water.

All engines use energy to make things move! But where do they get it?

Most engines, like those in cars, motorcycles, lawnmowers and boats, get their energy from fuel like gasoline.

Some engines run on electricity. Battery powered garden tools like leaf blowers and trimmers are examples.

For years people have even been working on making cars with engines that run on solar power—light from the sun.

Someday you may drive a car that doesn't use gasoline or electricity as its main source of energy, but uses energy directly from sunlight!

51

ENGINES THAT USE MOTION ENERGY

Some engines use energy from moving air or water. Wind and flowing streams provide a lot of energy naturally.

A windmill is an engine that captures wind energy to do work.

For more than a thousand years, people used windmills to grind grain into flour.

Grain was put between two large, flat, round stones called millstones. Power from the windmill turned one of them.

This crushed the grain between the two stones and turned it into flour.

ENGINES THAT USE STORED ENERGY

Most engines use some kind of stored energy.

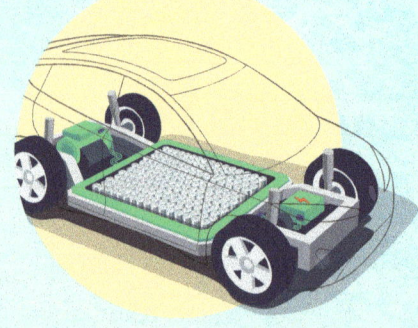

A lot of cars use energy stored in gasoline.

Some use electric energy stored in batteries.

Rockets use energy stored in rocket fuel.

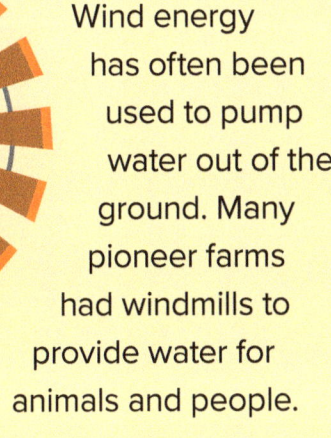

Wind energy has often been used to pump water out of the ground. Many pioneer farms had windmills to provide water for animals and people.

Another engine that has been used for thousands of years is a water wheel. This is a large wheel with paddles or blades. Running water, such as a stream or waterfall, pushed on the paddles and this turned the wheel.

Just like windmills, water wheels were often used to grind grain. They were also used in sawmills to cut huge logs into lumber.

Through a lot of human history, windmills and water wheels have been important engines that run on motion energy found in nature.

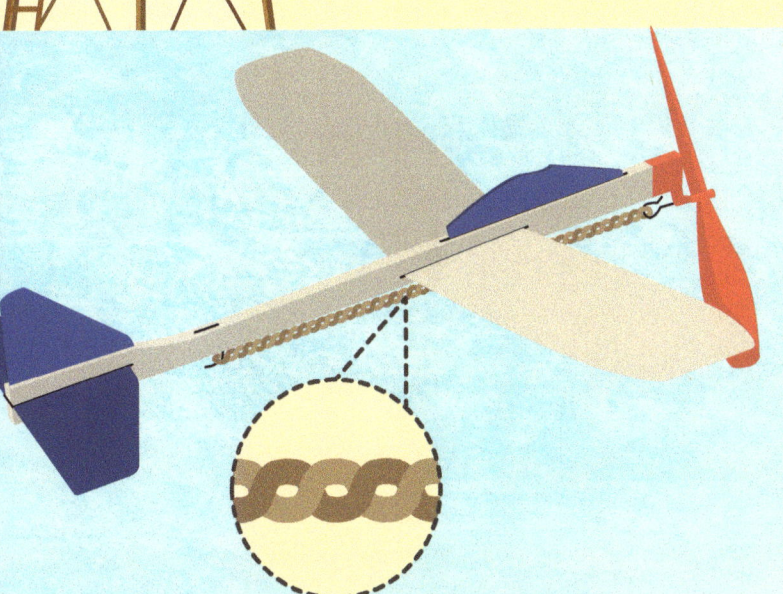

You can make or buy a simple toy airplane with an engine that runs on stored energy from a rubber band.

The rubber band is hooked to a propeller. You wind the propeller up by turning it with your finger. As you do this, the rubber band twists and stores up energy from the movement of your hand.

When you let go of the propeller, the rubber band quickly unwinds. This makes the propeller spin and it pulls the plane through the air. The rubber band is the airplane's engine.

You can make your own rubber band engine for a toy helicopter. To find out how, just turn the page!

RUBBER BAND HELICOPTER

For this activity you will need

- 1 rubber band helicopter kit
    or
  1 rubber band helicopter propeller
- 2 rubber bands – 3 ½ x ⅛ inch (size #33)
- 1 popsicle/craft stick
- 1 regular-size, plastic-wrapped paper clip
- file folder or similar stiff paper
- plain white paper
- tape

Steps

1. Fit the propeller onto one end of the craft stick.

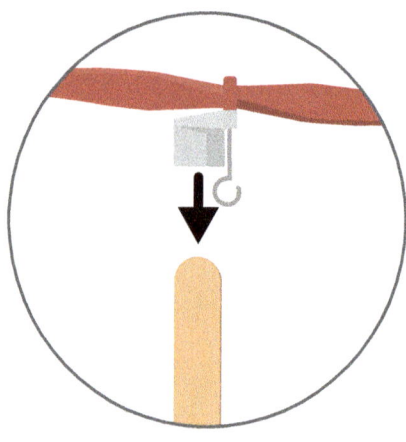

2. Unfold the paper clip until it looks like this

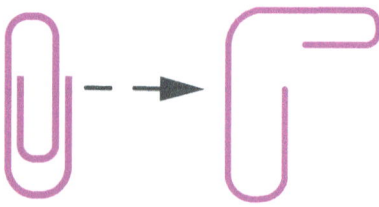

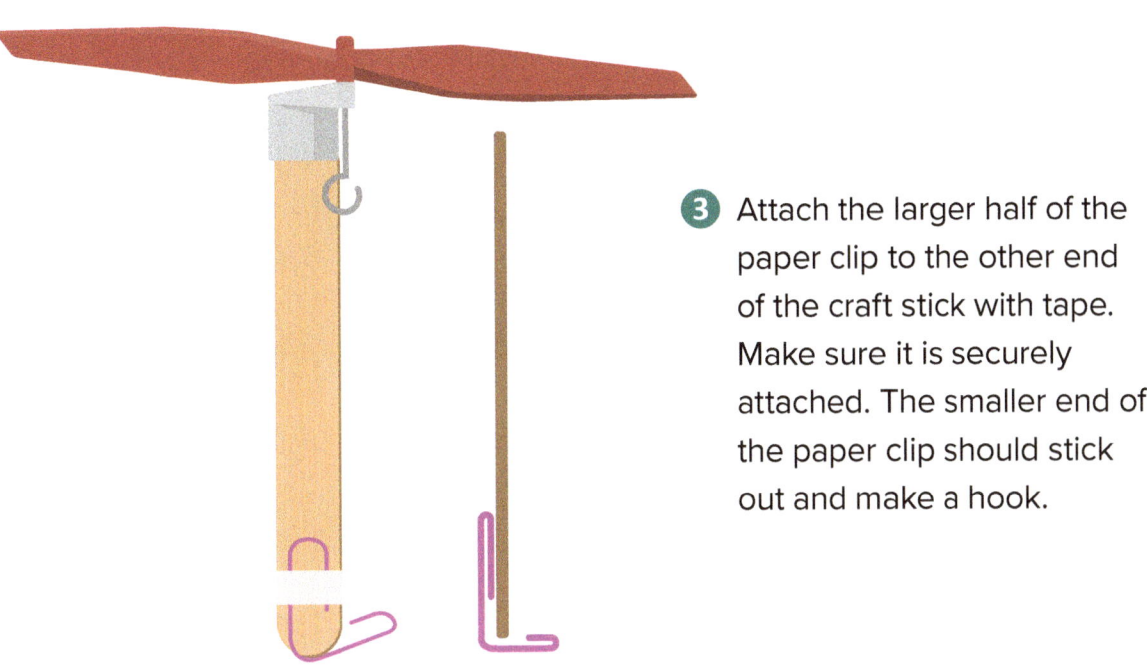

3. Attach the larger half of the paper clip to the other end of the craft stick with tape. Make sure it is securely attached. The smaller end of the paper clip should stick out and make a hook.

4. Trace the helicopter body onto plain white paper and cut it out.

5. Use your cut-out shape to trace the helicopter body onto a file folder or piece of stiff paper, then cut it out.

6. Tape the helicopter body to the craft stick near the propeller end. Make sure the helicopter body and the paper clip hook are on opposite sides of the craft stick.

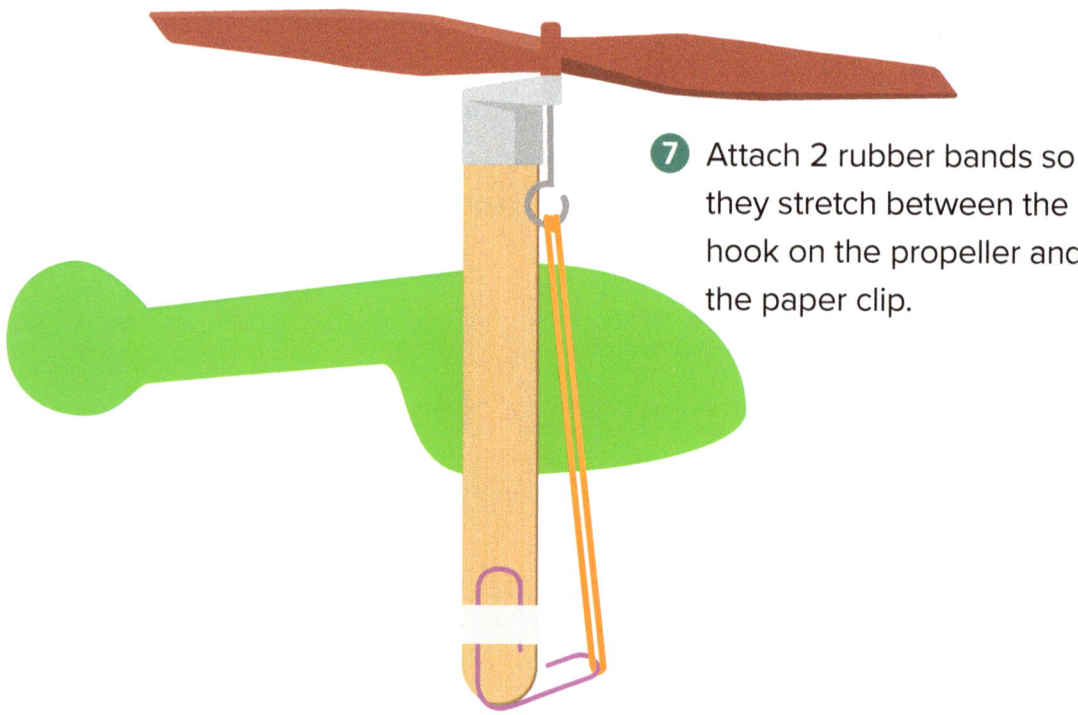

7. Attach 2 rubber bands so they stretch between the hook on the propeller and the paper clip.

8. Hold the paper clip and wind the propeller in a clockwise direction. Keep winding until the rubber band starts to coil tightly on itself.

9. Hold the top of the propeller in one hand and the bottom of the craft stick in the other.

⑩ Let go of the propeller first and then the craft stick. You can say "top, bottom" as you let go to help you remember.

⑪ Did it fly? How high?

⑫ (Optional) Experiment with different shapes or material for your helicopter to see if you can improve how high it goes or how well it flies.

⑬ Tell (or write up for) another person what makes your helicopter an engine, and explain what energy changes you observed in this activity and what happened.

11 CHAPTER ROCKETS AND JETS

If you want something to move very fast, a jet engine might be just what you are looking for!

A **jet** is a stream of liquid or gas that shoots out through a small opening at a high speed. A jet can create a strong force.

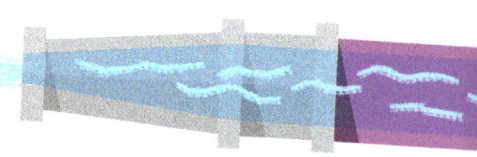

A **jet engine** is powered by a jet. It creates a jet of liquid or gas and shoots it out rapidly. A large jet engine can cause a push strong enough to help move an airplane at a high speed!

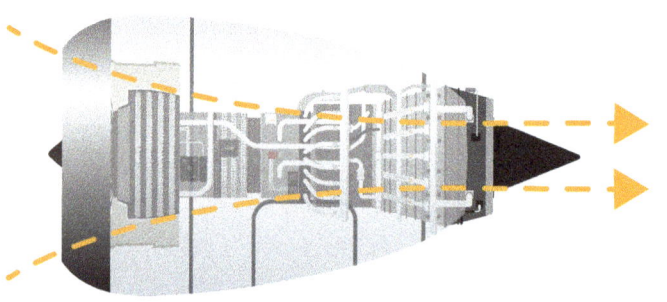

Jet engines can be very powerful. They are able to use a lot of force to move things very fast.

Fighter jets used by the Air Force can fly up to 3,000 miles per hour!

A jet engine can provide the power to make a jet airplane fly across the country at speeds over 500 miles per hour.

58

A SIMPLE JET ENGINE

Probably the simplest example of a jet engine is a balloon. When you blow it up with air, the motion energy of your breath stretches the rubber of the balloon. The balloon now has stored energy.

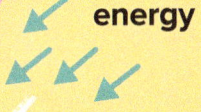

When you let go of the balloon, the stored energy pushes the air back out in a narrow stream or jet. As the air rushes out in one direction, the balloon moves in the other direction.

The balloon is a jet engine. If you attach it to a rocket body, you will have a rocket that can shoot off into the air.

A rocket needs to go 25,000 miles per hour just to break away from the gravity of Earth and reach outer space!

ROCKET ENGINES

A jet engine used to fly a rocket is called a *rocket jet engine* or simply a *rocket engine*. This engine burns fuel to create a very strong push, but it still works like a balloon jet.

Most jet engines use jets of gases. The helium pumped into a balloon that floats in the air is an example of a gas. So is the oxygen we breathe in from the air around us.

Burning fuel in a rocket engine produces gases and lots of heat. So much heat energy is created that the gases get extremely hot and start moving very fast. They shoot out of the engine with tremendous energy in one direction. This creates a huge push that sends the rocket in the other direction.

Putting out all that force takes a lot of stored energy. That means rockets need lots of fuel. In fact, they need so much fuel just to get off the ground that, before lift-off, almost 90% of a rocket's weight is fuel!

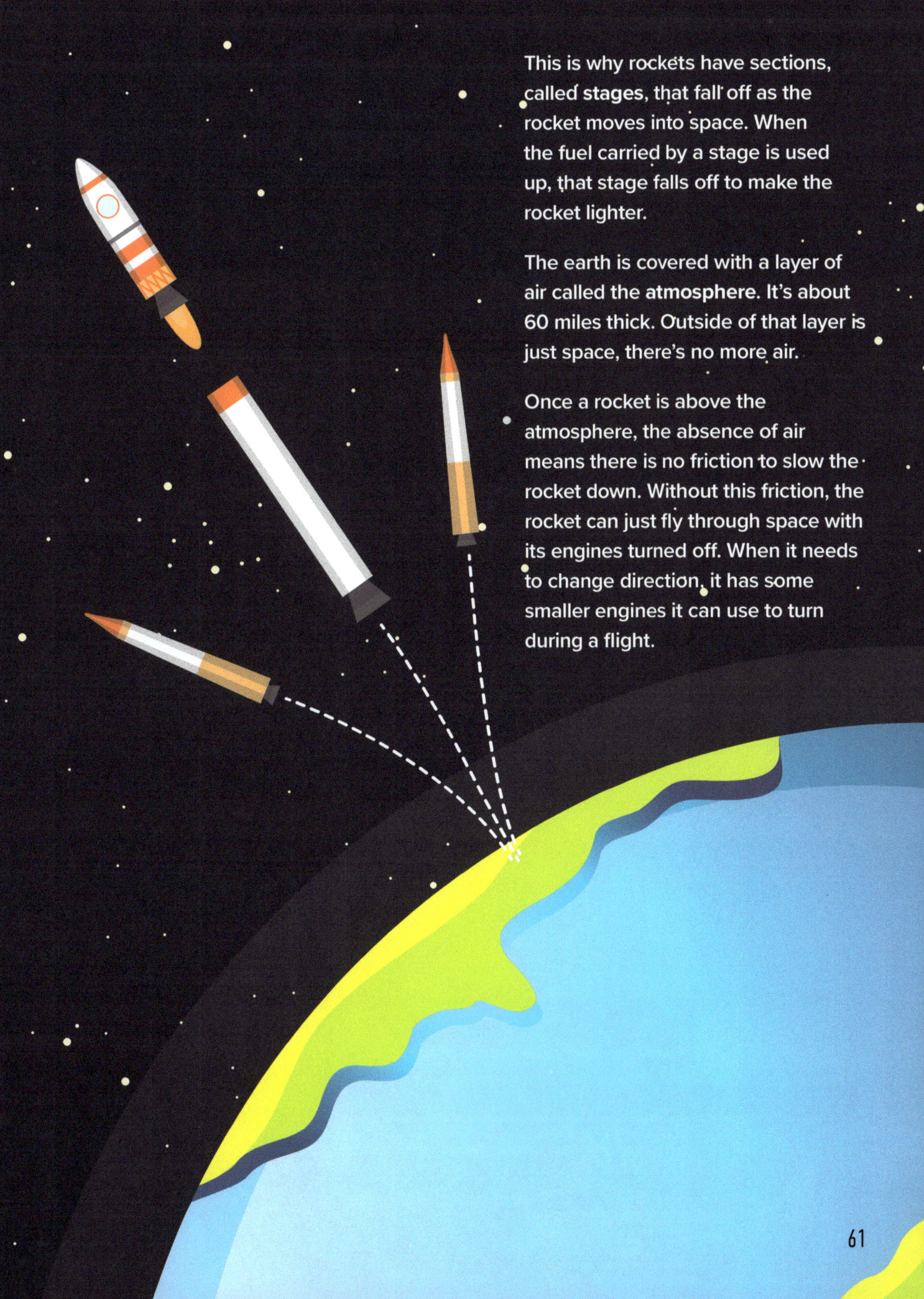

This is why rockets have sections, called **stages**, that fall off as the rocket moves into space. When the fuel carried by a stage is used up, that stage falls off to make the rocket lighter.

The earth is covered with a layer of air called the **atmosphere**. It's about 60 miles thick. Outside of that layer is just space, there's no more air.

Once a rocket is above the atmosphere, the absence of air means there is no friction to slow the rocket down. Without this friction, the rocket can just fly through space with its engines turned off. When it needs to change direction, it has some smaller engines it can use to turn during a flight.

JET AIRPLANE ENGINES

The engine of a jet airplane is a little different than a rocket engine.

In a jet airplane engine, outside air is sucked into the front of the engine. Then it is burned with fuel to produce the strong push needed to move the airplane.

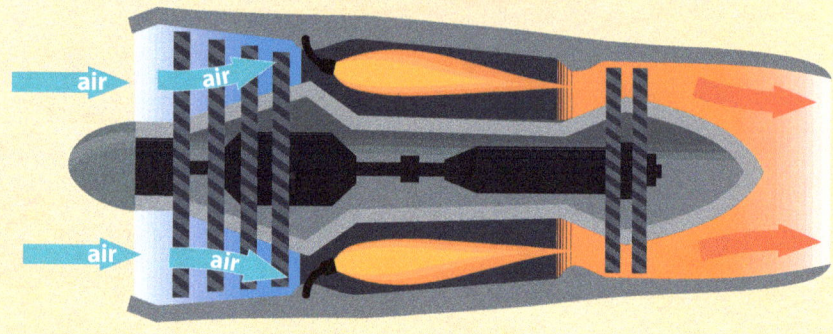

Parts of the engine keep the hot gases from coming back out of the front and only let them out the back.

Jet airplanes also carry a lot of fuel, but much less than a rocket.

One of the ways jet airplanes save on fuel is to fly high in the sky. There is still friction because the jet is only about 4 miles high and the atmosphere is 60 miles thick. But as you get higher in the sky, the air gets thinner, so there is less friction.

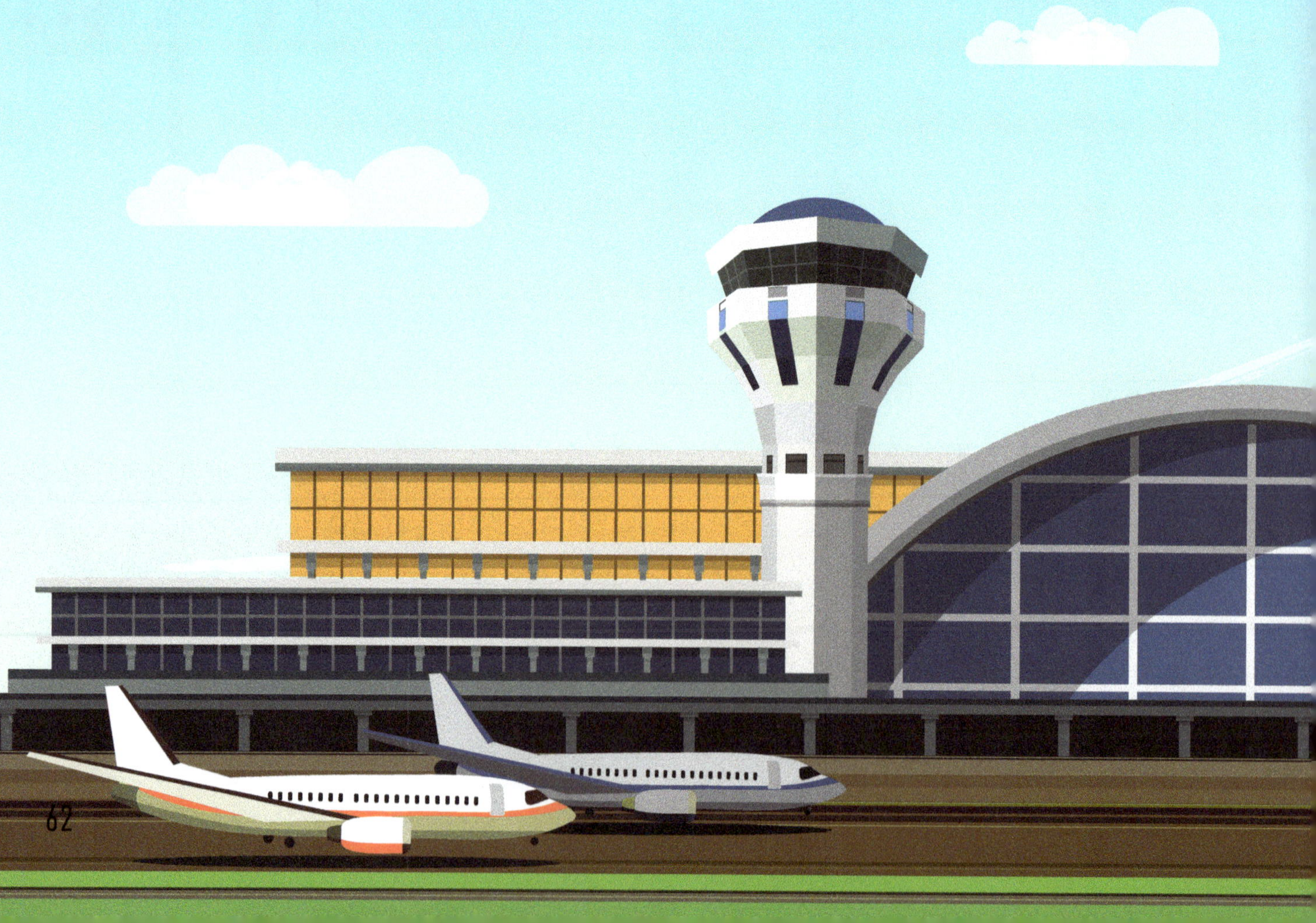

The hot gases shooting out the back of the engine push the plane forward.

BALLOON JET ENGINE

Here is a simple balloon jet engine that will go a long way.

For this activity you will need:

- 9-inch balloon
- piece of plastic straw 4 inches long
- yardstick
- scissors
- piece of thin cotton string at least 20 feet long
- 2 chairs
- tape

Steps

1. Position two chairs about 10 feet apart.

2. Thread one end of the string through the piece of plastic straw.

3. Tape one end of the string to the top of one of the chairs. Tape the other end of the string to the second chair. You'll have extra string but don't cut it off.

4. Move the chairs apart until the string is tight. Then slide the straw piece to one end of the string.

5. Blow up the balloon and squeeze or twist the open end to keep the air from escaping. Hold the balloon against the bottom of the straw with the open end of the balloon pointed toward the nearest chair.

6. Have someone help you tape the straw and balloon together at both ends of the straw. Don't bend the straw.

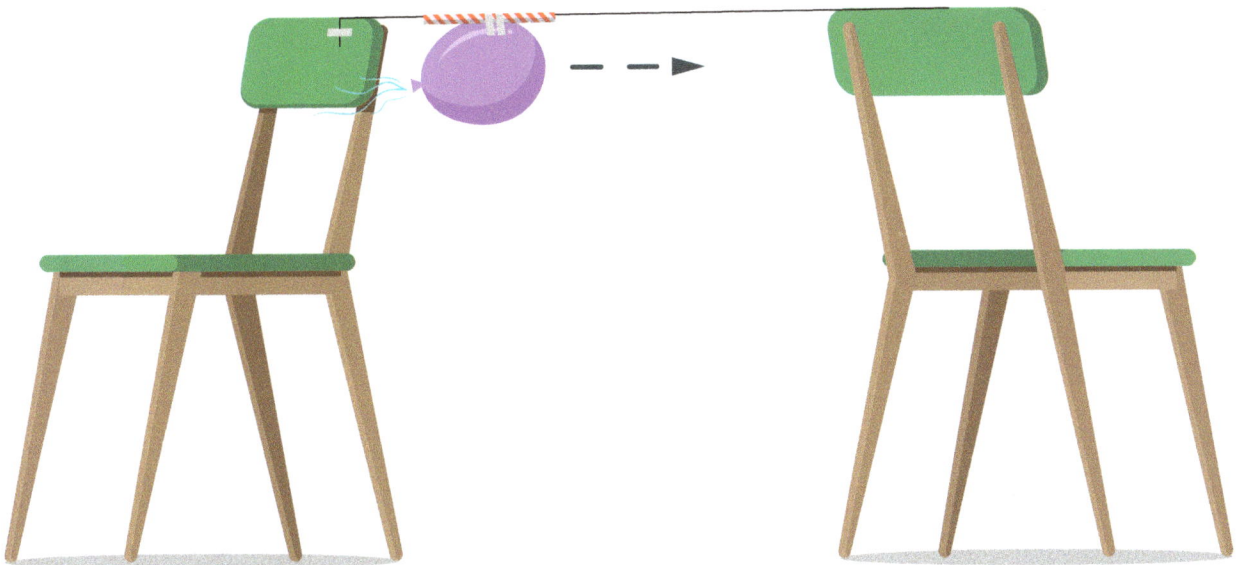

7. Release the balloon and watch what happens. The balloon should shoot all the way to the other chair.

8. Move the chairs 20 feet apart and repeat the activity. Did the balloon make it the whole distance?

9. Now see if your balloon jet is strong enough to fly upward against gravity for 20 feet. Keep one end of the string attached to a chair. Tape the other end of the string to the wall at least 3 feet higher than the end on the chair. Pull the chair back until the string is tight again. Repeat the activity and see what happens.

10. See how steep you can make the string before the balloon jet can't go to the end.

11. Tell (or write up for) another person how your balloon jet engine works and what you observed.

BUBBLE JET ENGINE

Gas bubbles can be used to push a machine the same way air and fuel do. This engine uses vinegar and baking soda to make bubbles.

For this activity you will need

- vinegar
- baking soda
- teaspoon
- several paper towel squares about 4 inches on each side
- scissors
- small, empty plastic dishwashing liquid bottle
- sink or bathtub

Steps

1. Fill a sink or bathtub with enough water for the bottle to float.

2. Open the pull-spout cap of the bottle all the way. Then unscrew the cap of the bottle and take it off.

3. Fill the bottle half full of vinegar. Don't put the cap back on yet.

④ Put one full teaspoon of baking soda into the center of each of your paper towel squares.

⑤ Roll up each square and twist the ends to keep it shut. Make sure each roll is narrow enough to go through the mouth of the bottle.

⑥ Push one of the rolls into the bottle. As fast as you can, screw the cap back on and put the bottle back into the water. Watch what happens. Did the bottle move?

⑦ Now experiment to see what a different size opening does to the bubble jet. Start by emptying the bottle and re-filling it halfway with fresh vinegar.

⑧ Close the pull-out cap all the way, then open it a tiny bit.

⑨ Push another baking soda roll into the bottle, screw the cap on again (quickly) and put the bottle back into the water.

⑩ Did it work differently this time?

⑪ (Optional) Experiment to see if you can improve how well your bubble jet engine works.

⑫ Tell (or write up for) another person how your bubble jet engine works and what you observed.

MATCH ROCKET ENGINE

For this activity you will need

- box of wooden kitchen matches
- wood skewer
- small candle
- aluminum foil
- scissors
- needle-nose pliers
- small piece of modeling clay

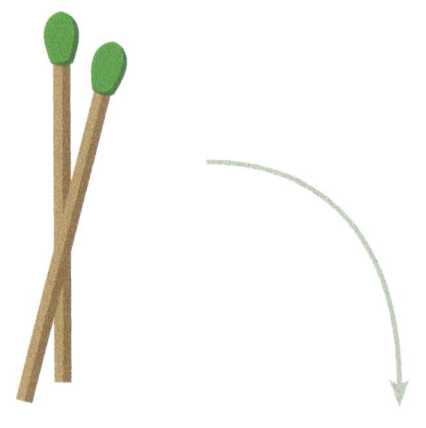

Steps

MAKE THE ROCKET

1. Use the pliers to cut the heads off 2 matches.

2. Cut the skewer in half. It should be 16 cm (about 6 inches) long.

3. Cut out a piece of aluminum foil like this:

④ Lay the flat end of skewer on the aluminum foil along the 5.5 cm (2-inch) side and put the 2 match heads at the tip of the skewer. They should be 1 cm (about ½ inch) from the top of the foil.

⑤ Roll the foil tightly around the skewer and the match heads.

⑥ Fold over the top and pinch tightly with pliers. This needs to be very well closed. It works best to pinch it into a round shape.

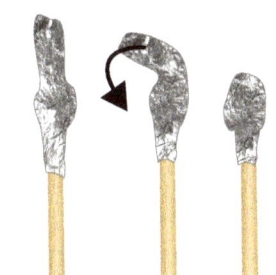

⑦ Pull the rocket off the skewer and set it aside.

MAKE THE LAUNCH PAD

① Empty the matchbox and punch a hole in the top of the box about 1 inch from the end. Put the inside section of the box back. This is your launch pad.

② Place the skewer in the hole with the pointed end out. Hold it in place with a small piece of clay.

③ Place the foil rocket back on the skewer. The rocket should be over the end of the launch pad. Be sure the rocket is pointed away from people and anything that could catch fire.

READY, SET, LAUNCH!

① Place a small candle under the tip of the rocket in a small piece of clay so the flame will touch the tip of the rocket. Light the candle and wait a few seconds to see what happens!

② Be careful when you touch the rocket. It will be hot!

③ Make three more match rockets and ignite them. Experiment to see if you can make them go farther or higher than the first one.

④ Tell (or write up for) another person how your match rocket engine works and what you observed.

CHAPTER 12: ELECTRIC MOTORS

Another name for an engine is a motor. When people are talking about motors, however, they very often mean electric motors.

An **electric motor** changes the electrical energy that comes from a battery or power station into motion energy.

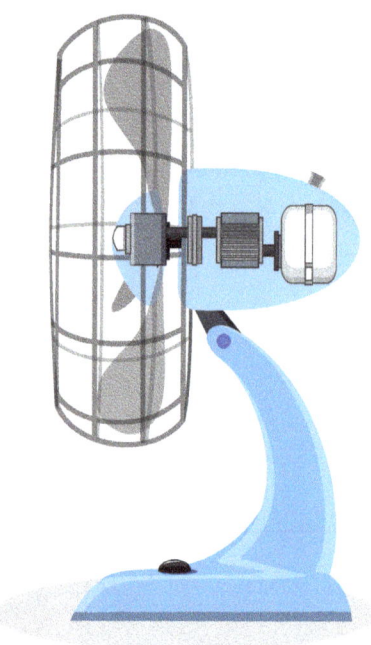

When you plug in an electric fan and turn it on, for example, a motor inside it turns electrical energy into the motion of the blades turning around.

Battery-powered toy cars have little electric motors in them that make the wheels turn.

A washing machine has a motor that makes the inside of the washer turn and get clothes clean.

The hum of a refrigerator? That's an electric motor.

The motor in the fan spins the blades.

motor

Blenders, vacuum cleaners and electric can openers all have electric motors. They make blenders spin, vacuum cleaners suck up dirt and can openers cut open cans.

An electric toothbrush has a tiny motor inside that makes the brush vibrate.

Today more and more full-size cars use electric motors to get around.

Computers have small electric motors inside that run fans to keep them cool. Even a smartphone has a tiny motor inside that makes it vibrate.

In all these devices, motors use electrical energy to produce motion. Because of motors, they can all do their work.

TABLE FAN

For this activity you will need

- 1 rubber band helicopter propeller
- 6-volt motor
- 6-volt lantern battery
- 2 alligator clip wires
- cardboard toilet paper tube
- piece of corrugated cardboard (4" wide and 8" long)
- scissors
- needle-nose pliers
- hot glue gun

Steps

1. Trace the end of the toilet paper tube on one end of the cardboard and cut out the circle.

2. Use hot glue to attach the circle to one end of the toilet paper tube.

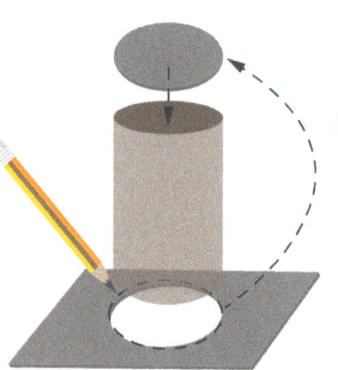

3. Glue the other end of the toilet paper tube to the remaining cardboard. This is the stand for your fan.

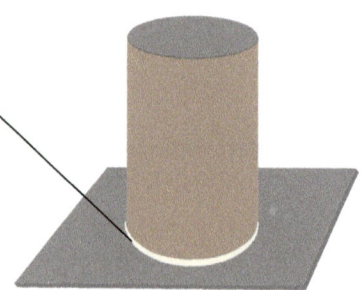

④ Use pliers to straighten the wire that goes through the center of the propeller. Slide the wire free.

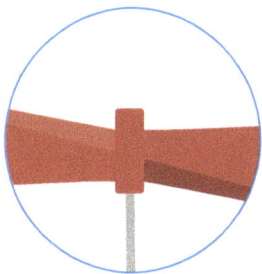

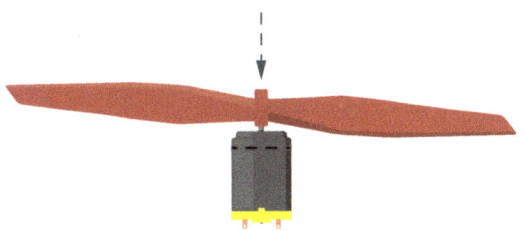

⑤ Push the propeller onto the shaft of the 6-volt motor.

⑥ Use hot glue to attach the motor to the top of the toilet paper tube.

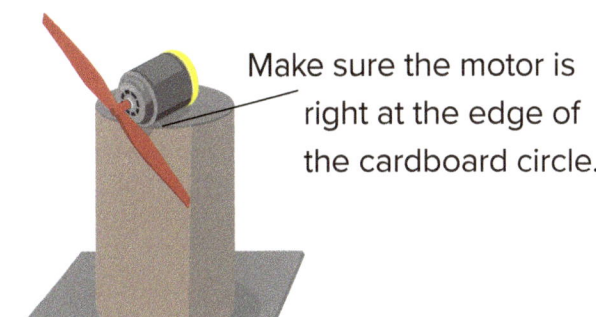

Make sure the motor is right at the edge of the cardboard circle.

⑦ Connect the motor to the 6-volt battery with the alligator clip wires and see what happens.

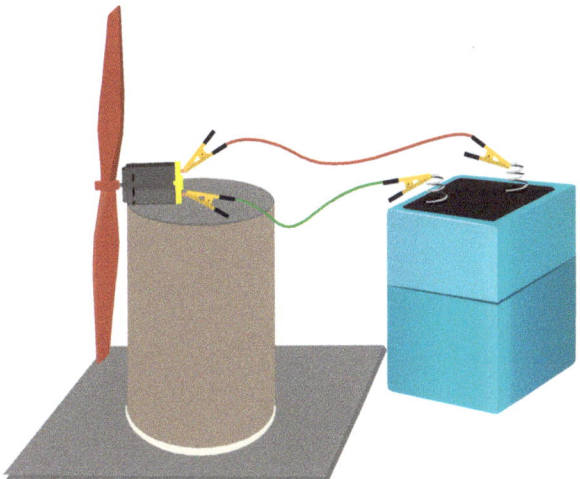

⑧ If the fan blows in the wrong direction, just switch the position of the wires going into the motor.

⑨ Tell (or write up for) someone what you did and what you observed.

13 CHAPTER SOME INTERESTING

Engines give power to many machines around us and make them run. They make it possible for us to travel in cars, build huge skyscrapers, carry heavy loads on ships, travel through the air, even go to the moon and Mars.

Though they all make things go, engines come in a wide variety of shapes and sizes.

One of the largest engines in the world is the size of a small apartment! It's 87 feet long and 44 feet high. This engine is used to power huge oceangoing cargo ships at speeds higher than any earlier ship engine.

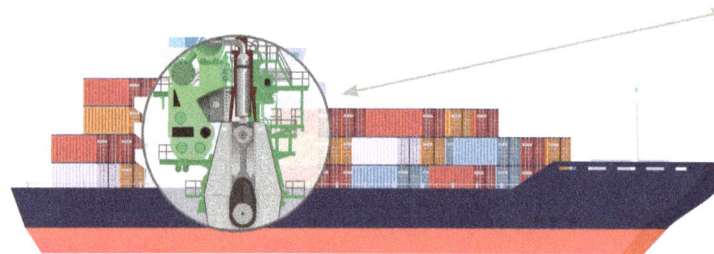

Some of the smallest engines in the world are about the size of a penny. They are used to run remote control cars. If you've ever played with remote control cars, you know from experience these little engines can really make things move!

74

ENGINES AND MACHINES

The way to measure the amount of power an engine puts out is with **horsepower.** This word horsepower was first used back when steam engines were being invented. People still used horses for work, and they wanted to know how powerful the new steam engines were. It made sense to compare how much work a steam engine could do to how much a horse could do. If a steam engine could do the work of four strong horses, it was said to have 4 horsepower.

The engine of a riding lawnmower has about 15 horsepower.

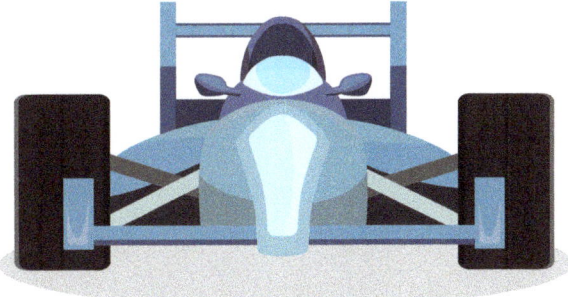

A race car has about 750 horsepower.

The huge cargo ship engine has 107,389 horsepower. That's an engine that can really move things!

75

MARS ROVER

NASA is a group of American scientists who study outer space by designing and building rockets. They also plan and carry out space exploration missions.

On the next page is a picture of NASA's Perseverance Mars rover. ("Perseverance" means sticking with something even though it's hard. It's easy to see why a Mars rover ended up with this name!) This is a machine that was created to move around Mars to find out more about its climate and what its surface is like. The Perseverance rover is about the size of a car. It has hundreds of motors that move all its different parts.

It has an arm that can drill into rock to get samples.

It's hard to imagine all of the people who studied and learned about engines so that we could land something like that on Mars. It's doing work for us 160 million miles away from Earth!

It even has its own mini-helicopter.

It has 19 cameras for taking pictures and 2 microphones for recording sounds.

The world is full of fascinating machines, motors and powerful engines that make things go! If you want to know more about engines, how they work and what they do for people...

Well, you're a young scientist.
Go find out!

PULSE JET ENGINE

For this activity you will need

- pop pop boat kit
- candle
- matches
- sink or large basin
- small medicine dropper

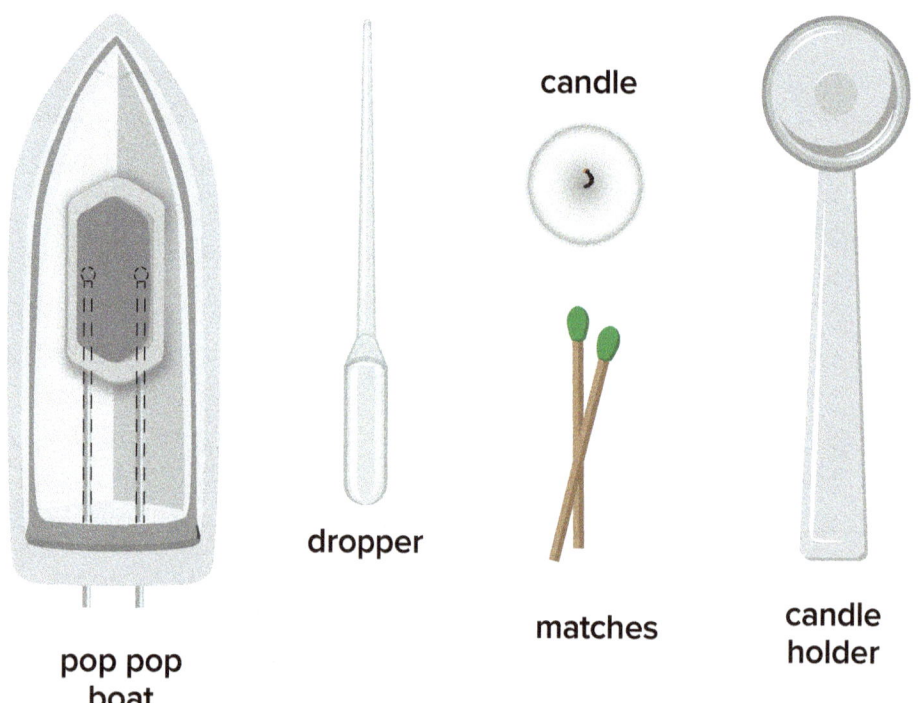

pop pop boat · dropper · matches · candle holder

A **pulse** is a spurt that repeats again and again. This engine is called a pulse jet engine because it squirts out water again and again, perhaps 5 to 10 times a second.

Steps

1. Fill a sink or large basin with water.

2. The engine of a pop pop boat runs on steam. Find the boiler and tubes on your boat. (A **boiler** is a tank in which water is boiled to produce steam.)

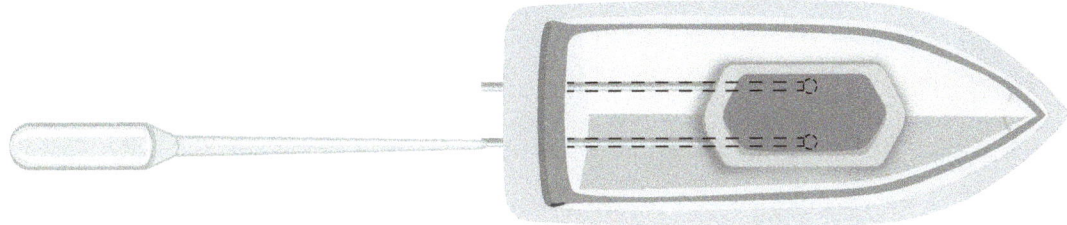

3. Use the dropper to put a few drops of water in one of the tubes at the back of the boat. Keep adding water until it runs out of the other tube. This will fill the boiler.

4. Put the boat in the water and put the candle on the floor of the boat.

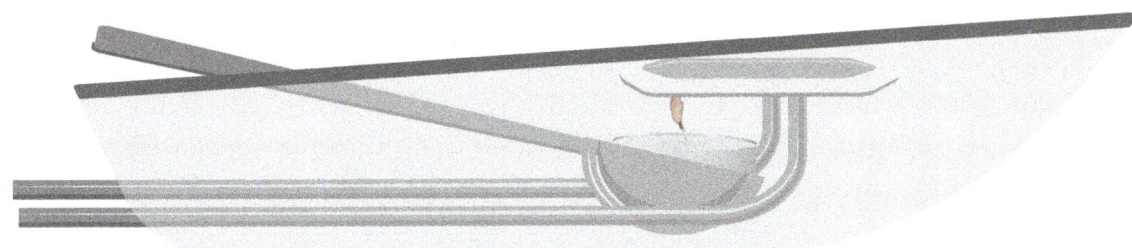

5. Light the candle and slide it under the boiler.

6. After a few seconds the boat will start moving. See if you can tell why it is called a "pop pop" boat.

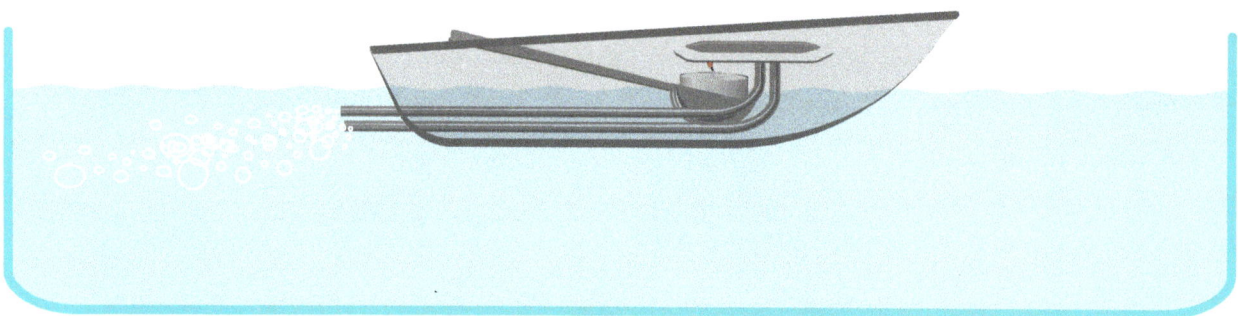

7. Let the boat go for about a minute, then blow out the candle. See how long it takes for the boat to stop. **Note:** The front end of the boat will be very hot. Let it cool off before you take it out of the water.

How does it work?

- The engine must start with some water in the boiler.
- Heat from fuel (a candle) boils the water and turns it to steam.
- The steam pushes out the water in the tubes in a pulse. The water shoots out through two small tubes in the back and pushes the boat forward.
- The steam left in the boiler cools off and takes less space. That lets water from below the boat push back into the boiler.
- The candle heats the steam again and pushes out the water again, and again the boat moves forward. These steps can happen very fast, perhaps 5 to 10 times each second, so there might be 5 to 10 pulses of water each second.
- The steam continues to heat (pushing out water) and cool (sucking in water) again and again as long as the candle burns.

8. Tell (or write up for) someone what you did and what you observed.